W9-CGV-548

BASIC TELEVISION

Vol. 1

BASIC TELEVISION

Revised Second Edition Vol. 1

ALEXANDER SCHURE

HAYDEN BOOK COMPANY, INC.

Rochelle Park, New Jersey

ISBN 0-8104-5060-7
Library of Congress Catalog Card Number 74-2957

Hayden Book Company, Inc.
50 Essex Street, Rochelle Park, New Jersey 07662

Printed in the United States of America

1 2 3 4 5 6 7 8 9   PRINTING

74 75 76 77 78 79 80 81 82   YEAR

PREFACE

Fifteen years have passed since the initial publication of the five-volumed course dealing with basic principles of television. During these years we have seen extraordinary social and technological changes, ones which are now touching all aspects of our society. It seems more than likely that in the future these changes will evolve at an even more rapid rate. The world of the twenty-first century will be as different from that of today as our present world is from that which existed prior to the industrial revolution.

With each passing day we witness developments that are continually breaking with the past. In this last decade alone, we have moved from jet travel to space travel, from limitations of travel within the atmosphere of earth to the capacity to probe Mars and beyond. Electronics, too, has progressed with similar speed. Color television, facsimile, computers, transistors, and LSI memories are all recent products.

The present six-volumed course represents substantial modifications of the original basic television programs. In addition, particular emphasis has been given to the development of color TV and the introduction of new solid-state approaches. The heavily illustrated format adapted throughout the initial volumes has been retained. The combination of text and illustrations has been chosen in order to place particular emphasis on essential materials, and, in addition, review sections are interspersed throughout each volume in an effort to capsulize major points covered in the body of the text. As a result, the overall visual approach and programmed sequence coverage is comprehensive, from the creation of the television image in the studio to its appearance on the home screen, and presupposes only a knowledge of basic electronics.

The author wishes to acknowledge the substantial contributions of Messrs. Harvey Pollack, Charles J. Anderson, Jr., Fred R. Kulis, and Anthony S. Santonelli in the updating of these volumes.

Alexander Schure

Old Westbury, New York

CONTENTS

Vol. 1

Television Broadcasting

The universe at our fingertips; that's what the magic of television has made possible for all of us. We can view the surface of the moon just as clearly as we can see our favorite team score that critical goal. It is no wonder, then, that TV is attracting the widest audience in the history of entertainment and communication.

The future of television broadcasting is limitless, with its expansion dependent only upon the creative genius and perception of man. Whatever is visible, whether in the deepest recesses of the human body or in the farthest regions of outer space, can be photographed by the TV camera and broadcast in an instant throughout the world. Not only is man entertained and informed by television, but also the multiuse of TV by men of science, medicine, industry, and education can expand man's horizons to unimagined proportions.

The Television Team

To fully understand what happens inside a television receiver, you should first know what goes on at the transmitting end: how the TV camera picks up a scene which is then converted to electrical signals; how sounds happening at the scene are mixed with the picture signals to form composite signals; and finally, how the composite signals are broadcast by the TV transmitting antenna.

At the receiving end of the system, the receiving antenna pulls the signals out of the air and feeds them into the TV receiver where they are converted back to picture and sound. To perform this electronic miracle, it takes a team of highly skilled professionals working closely together.

It Takes a Team for...

TODAY'S TV LINEUP

1. Script Writer
2. Actors
3. Director
4. Technical Director
5. TV Cameramen
6. Announcers
7. Lighting Technicians
8. Stagehands
9. Video Engineers
10. Audio Engineers
11. Transmitter Engineers

TV BROADCASTING

Television broadcasting not only requires the services of many different skilled people, but even more important, it requires teamwork and accurate timing. To achieve this takes more than just rehearsals. From beginning to end the electronic equipment must function perfectly. Program sponsors set very high standards, and the public now expects perfection. Defects in equipment along the transmitting chain must be identified and corrected immediately so that the TV viewer receives perfect picture and sound.

We shall begin our course in Basic Television, therefore, with an in-depth look at what happens at the transmitting end of the television broadcasting system.

The Television Studio

With the show in production there is little room for mistakes. A cue missed by a sound-effects man can easily destroy an illusion, a slip by a director can destroy days of careful rehearsal, a wrong twist of a control room dial can cause expensive retaping. Even though magnetic video tape recording is used for most studio-produced programs, there is still no margin for error; perfection is always the goal just as if the program were being shown live. Frequently the TV actors give better performances because the most intolerable pressures of live TV have been relieved through the use of video tape recording.

ON STAGE IN THE TELEVISION STUDIO

The work of the television show as a whole is guided by one man, the *director*. Not only does he utilize the services of engineers, technicians and cameramen, but he also directs their operations. Once he has started to record the show on video tape, he tries to continue without a flaw. If retakes are required because of flaws he has detected while the show is being recorded, he directs the retakes from his vantage point in the studio control room. From here he watches the show through a glass partition. It is from this control room that the director unveils his talent to the viewing audience.

The Television Camera

All television programs begin with the camera, thus making the camera the most important element in the system. All other elements in the system are set up to take full advantage of the camera's characteristics. For most programs, whether outdoors or in the studio, as many as three or four cameras may be used, with each one delivering a different view to the director in the control room. There the director selects the scene to be televised.

THE TELEVISION CAMERA

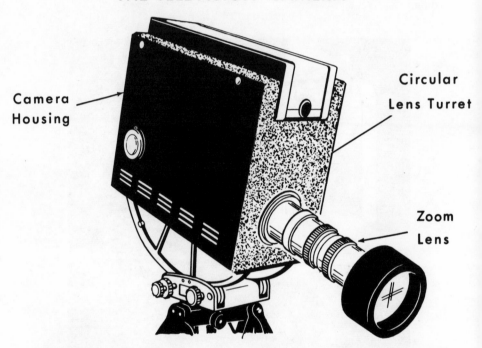

Camera Housing

Circular Lens Turret

Zoom Lens

On older TV cameras four different lenses were often mounted on a circular turret. The lenses gather the light from the scene and focus it as a small, sharply defined picture on a specially prepared surface inside the camera tube. Only one lens works at any one time. By rotating the lens turret the cameraman brings into action any one of the lenses. The long lens is a *tele-photo* lens generally used for closeups. The other lenses permit the camera to get normal shots and closeups, depending on the distance to the subject.

The camera housing encloses the camera tube and the electronic equipment directly associated with it. The entire camera is mounted on a "dolly" which can be rolled along the floor as desired. The handle permits panning, that is, moving the camera both vertically and horizontally.

The Television Camera (cont'd)

In modern television systems, two types of cameras are used: the *monochrome* or black and white camera, and the color camera. Both types operate on the same basic principle: the camera converts what it sees into electrical signals. In this volume we discuss the monochrome camera principally. In today's modern monochrome cameras, three main types of camera tubes are used. One type uses the *image orthicon;* another smaller camera uses the *vidicon;* the newest camera uses the *plumbicon.*°

Essential Elements of the Television Camera

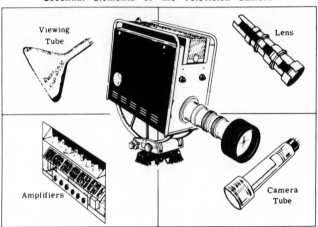

In general, a television camera consists of four major sections. One of these, the optical system, starts with what is probably the most critical item of all—the lens. In older cameras, several lenses on a rotating turret were almost instantly available to the cameraman for individual use as the need arose. Depending on the requirements of the scene, a lens was selected to focus the image onto the specially prepared surface of the camera tube. The modern camera, however, is equipped with a zoom lens which permits the cameraman to vary focal length over a 10:1 or greater range, going from a long-distance shot to a closeup smoothly and clearly. The second portion is the camera tube that translates light energy into electrical energy. Third are the amplifiers required for correct functioning of the camera tube and for amplification of the picture signal (video) output from the camera before it is delivered to other amplifiers outside the camera. The fourth portion of the studio-type camera is the viewing system which enables the operator to see the image picked up by his camera. It consists of a picture tube similar to, but smaller than, those used in the home, and its associated amplifiers.

° *Plumbicon* is a registered trademark by Philips Gloeilampenfabricken, Eindhoven, Netherlands.

Using the Camera

Let's examine a few tricks of the television broadcasting trade. They involve the television camera, manipulation of electronic circuits, and mechanical gimmicks. Many television techniques have been borrowed from the moving picture industry, including the manipulation of prisms in front of the camera lens for multiple-image effects. Many trick television images are accomplished by a change in the direction of electric currents which actuate certain portions of the camera.

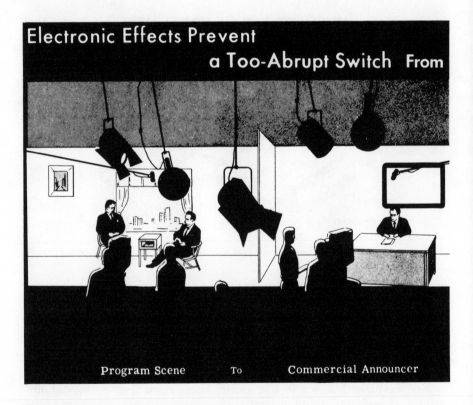

Electronic Effects Prevent a Too-Abrupt Switch From

Program Scene   To   Commercial Announcer

A stage-wise director demands many special effects from the camera. For example, he'll try to avoid an irritatingly abrupt switch from program scene to commercial. For a smooth transition he might use a fade—gradually fade the program scene into darkness and replace it slowly with the image of an announcer delivering his commercial. This procedure is at the command of the director, but is accomplished by a control operator. Both are stationed in the studio control room (which we describe later) from where the electronic behavior of the cameras is regulated.

Using the Camera (cont'd)

The Lap Dissolve

Another trick, employing two television cameras, is the lap dissolve. In it one scene is slowly faded out while a second scene is gradually brought into view so that for a short while the *two pictures overlap*. This technique, which was borrowed from the movies, is used to indicate that only a short time has elapsed between the actions associated with the two scenes. The fading scene, for example may show a man and woman seated in a railroad car. It is slowly replaced by a view of a hotel lobby, and after a moment a man and woman are seen walking into the lobby to the registration desk. The viewer at home immediately understands that the actions of leaving the train and going to the hotel have taken place.

For this effect the behind-the-scenes camera work is simple. With one camera trained on the railroad car set and a second trained on the adjacent hotel set, the director and monitoring engineers in the control room fade out the first picture and bring up the second, while the actors walk from the first set to the second.

Using the Camera (cont'd)

The Montage

A bit of electronic sleight-of-hand can be interesting in a television show. Youngsters viewing at home are enthralled when they see a giant taller than a mountain. The apparent magic is the result of a special montage effect.

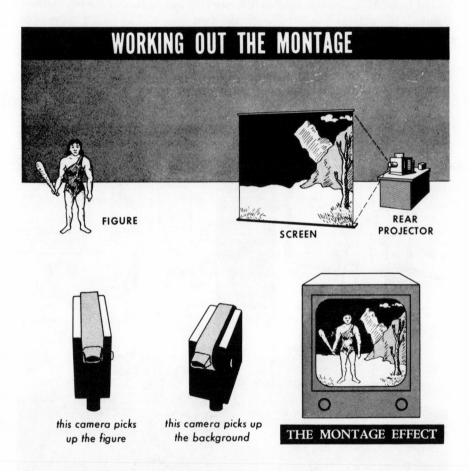

WORKING OUT THE MONTAGE

FIGURE

SCREEN

REAR PROJECTOR

this camera picks up the figure

this camera picks up the background

THE MONTAGE EFFECT

The mountain looks very realistic but it is actually a lantern slide whose image content is thrown onto a screen by rear projection. The man is real, dressed in a costume to suit the story. To perform this trick the outputs of two cameras are put on the air simultaneously; one camera picks up the lantern-slide scene and the other camera picks up an image of the man. The two views are combined into a composite signal and sent to the receivers. When viewed on the receiver the two pictures appear as one. The same result is also obtained when the actor performs in front of the lantern slide.

Using the Camera (cont'd)

Multiple Images

There are times when the effect of multiple images is desired. A simple method of accomplishing this is by mounting on the camera lens a prism that can be rotated around an axis. Almost any number of images can be obtained by setting the proper prism in front of the lens. When the prism is rotated the images revolve about one another on the television screen.

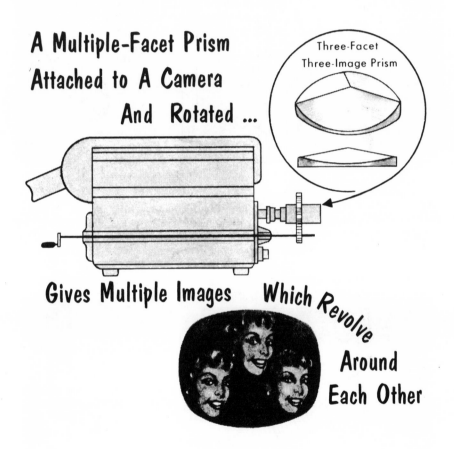

The triple image shown in the illustration utilizes a triple-facet triple-image prism. If eight images are desired an eight-facet eight-image prism is used. The number of images is determined by the number of facets on the prism. The television director has a bag full of tricks, but we shall not describe any more of them at this point; for now we want to probe the electronic details of the equipment.

Using the Camera (cont'd)

Many of the illusions seen on the television receiver screen are created with mechanical aids. They are utilized not only by television but also by the moving picture industry.

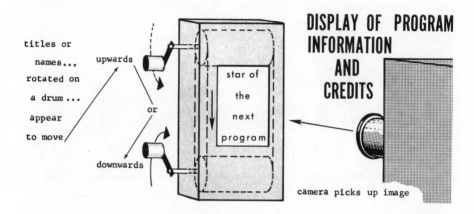

titles or
names...
rotated on
a drum...
appear
to move

upwards

or

downwards

star of
the
next
program

DISPLAY OF PROGRAM
INFORMATION
AND
CREDITS

camera picks up image

One special effect is produced by rotating a disc on which a spiral has been drawn. The rotation of this disc in a counterclockwise direction would make the spiral appear to run out of the receiver screen; rotating it clockwise would make it appear to be spiraling towards the center of the screen.

The "Running Line" Device

Inside the Television Camera

What is inside the housing of a television camera? Let us answer this question by examining two different kinds of cameras: the image orthicon monochrome (black-and-white) camera, and the vidicon camera.

Examining an exposed view of the image orthicon camera we see transistors and other electronic components together with a variety of controls—but no film. This is the feature which differentiates the TV from the movie camera.

The zoom lens shown allows the operator to vary the view presented to the TV audience from long distance shots to closeups with a kind of smoothness that was not possible when turret lens systems were in common use. The operator's viewing monitor, seen in the upper right-hand corner of the camera, is a standard picture tube that is the same, only smaller, as those used in home TV sets. The picture picked up by the lens and projected onto the light-sensitive surface of the image orthicon camera tube is reproduced on the monitor screen, so the operator sees what the camera picks up.

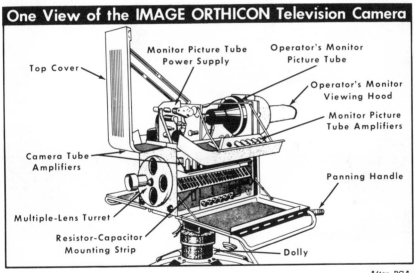

One View of the IMAGE ORTHICON Television Camera

Top Cover
Monitor Picture Tube Power Supply
Operator's Monitor Picture Tube
Operator's Monitor Viewing Hood
Monitor Picture Tube Amplifiers
Camera Tube Amplifiers
Panning Handle
Multiple-Lens Turret
Resistor-Capacitor Mounting Strip
Dolly

After RCA

The image orthicon camera tube is not visible in the drawing, but is behind the assembly of components which appear on the vertical band in line with the active lens. Operating voltages required for the functioning of the camera system and amplifiers are fed from outside the camera by special cables (not shown). A variety of amplifiers required for the operation of the camera tube, the viewing monitor picture tube, and the initial amplification of the camera-tube picture-signal output, are also part of the camera.

Inside the Television Camera (cont'd)

The vidicon camera is very much smaller and very much simpler than the image orthicon. Although originally intended for industrial and closed-circuit TV, the refinements made in the vidicon camera have widened its utility. Its portability makes it popular for field pickup, and it is enjoying increasing use in the television studio.

Our illustration shows one version of the vidicon. A complete outline of the vidicon camera tube is not visible in the drawing but it can be visualized as a long and narrow glass envelope inside the tube marked "light barrel." The single lens attached to the housing projects the image onto the photo-sensitive surface of the vidicon. On other vidicon cameras a high-ratio zoom lens is used instead of the fixed lens.

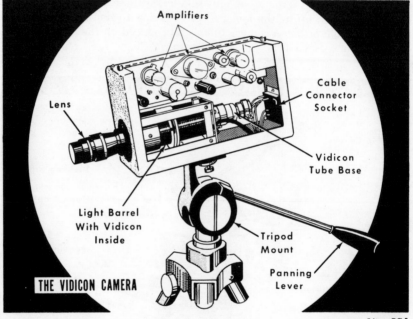

After RCA

The shielded tubes shown above the camera-tube portion of the camera are the video amplifiers in this particular model. More recent cameras utilize transistors and other solid-state electronic components instead of tubes. They amplify the video output before it is delivered to amplifiers outside the camera. The power supply which operates the camera tube is located in a separate unit not shown in the illustration. A cable connects the camera with the power supply and delivers the necessary operating voltages to the camera.

We have shown a manually controlled camera, as indicated by the panning lever attached to the upper tripod. Some cameras are hand-held. Also, some vidicon cameras can be remotely controlled.

The Studio

Have you ever been inside a television studio? Some are small, some large, and when they are large they are really big—75 to 100 ft long, 30 to 50 ft wide, with ceilings 30 or 40 ft overhead—large enough to put on a circus show and still not occupy the entire place.

THE MANY ACTIVITIES THAT GO ON INSIDE A TELEVISION STUDIO

Around the sides of the room 15 to 20 ft above the floor is a catwalk with a railing all around it. A maze of pipes criss-cross on the ceiling. From it hangs an assortment of lights, some fixed in location, others movable. There is usually ample room to put up three, four or more sets for the same show or for shows that follow each other. Even then there is room available. Scattered around the studio are cameras and microphones. Cables from these devices sprawl all over the floor and banks of lights illuminate one area. Every job is clearly planned, for the technique of broadcasting requires the most perfect timing. Any sound is picked up by the microphones while a show is on, so signals are given by means of printed signs, intercom microphones and head sets, and motions which comprise a sign language all of their own. Considering what goes on behind the scenes in the production of a television show, the degree of perfection of the programs one sees on a receiver screen is really astounding.

The Studio Control Room

Whether a program is being broadcast live or is being video-taped, the control room is truly the command post of the television broadcasting activity of any station. It is a specially designed room which usually provides a clear view of the stage. The director sits here among the studio control operators and master-minds the complete telecast. In front of him is a master switching panel with which he can select from the camera pickups displayed on the camera monitors the one scene he wants to transmit live or record on video tape.

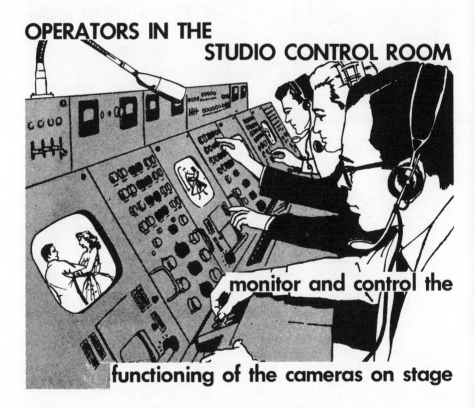

OPERATORS IN THE
STUDIO CONTROL ROOM

monitor and control the

functioning of the cameras on stage

Each control operator has before him the monitor which displays the scene picked up by the studio camera that is his responsibility. Manipulation of the "live" camera on the stage is the responsibility of the cameraman; manipulation of the electronic system within the camera is the function of the studio control operator. The scene is fed from the camera to the monitor by a cable and a related video amplifier located in the control room. The view of the control room shown here is a typical one. The number of control operators and monitors depends upon the size of the installation.

The Studio Control Room (cont'd)

Also located in the studio control room are the "air" and "preview" monitors, shown here in another studio control room. They are picture tube display systems to which picture signals are piped directly, rather than transmitted over the air. The air monitor displays the scene that the director has selected for transmission to the public. The preview monitor (or monitors) displays pictures that have been made ready to be put on the air; for example, a commercial, or some picture that is to be cut into the sequence of the video presentation. Of course, whatever sound must accompany the scene displayed on the preview monitor is transmitted when the preview monitor image is put on the air and appears on the air monitor.

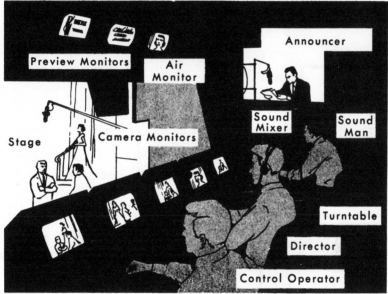

Another Studio Control Room

The studio control room also accommodates the *sound man*. He, too, takes his cues from the director. He monitors the sound that accompanies the action on stage, the announcer's comments, or whatever audio must be transmitted to the viewers. In front of the sound man is the sound mixer panel to which the cables are fed from the live microphones on the stage. The output from each microphone is controlled separately on the mixer panel, enabling the sound man to raise or lower the individual outputs as required, or to blend the sounds if necessary. When the sound accompaniment to a picture is delivered from several sources, it is made into a composite audio signal within the mixer, ready for delivery to other audio amplifiers and eventually to the transmitter.

The Studio Control Room Picture Chain

To clarify the relationship between the studio and the studio control room, wherein lie the duties of the director, we show a greatly simplified block diagram of the picture signal chain between these two parts of the television station. Many elements of equipment are omitted. They appear later as part of more detailed explanations.

In the diagram we have arbitrarily assigned certain duties to the three cameras on stage. Camera 1 is being used for full-length shots, camera 2 for closeups, camera 3 is mounted on an elevated dolly for overhead shots. These functions are indicated by the nature of the scenes being shot.

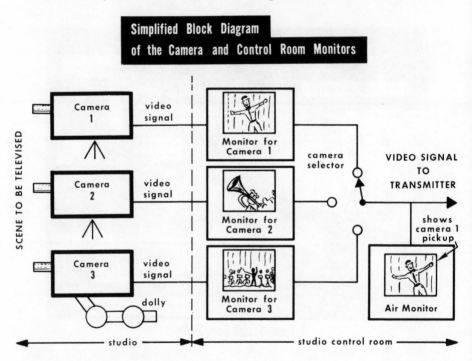

The three-position camera-selector switch is a simplification of the control with which the director selects the scene to be transmitted. We show only three positions because only three cameras are shown. A switching panel allows the director to issue instructions to the individual cameraman, microphone men, and the sound man in the studio. Cables that connect the director with the personnel on stage are not illustrated in the block diagram. The video cables, which feed the picture signal from the cameras to their related monitors, also contain conductors which supply the electrical energy to each camera. In addition, they permit voice communication between the men on the stage and in the studio control room.

1. *Production.* A television show can be successful only when all of the people and elements involved in the production are functioning with perfect coordination.

WORKING OUT THE MONTAGE

FIGURE · SCREEN · REAR PROJECTOR

this camera picks up the figure · this camera picks up the background · THE MONTAGE EFFECT

2. *Techniques.* Much of the effectiveness of television results from judicious use of electronic "tricks" performed with the camera and/or lenses, and "sleights of hand" involving mechanical aids.

The "Running Line" Device

THE TELEVISION CAMERA

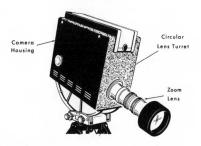

Camera Housing · Circular Lens Turret · Zoom Lens

3. *The Lens* is the beginning of the action of the camera, as the camera is the beginning of the televised scene. Many cameras are fitted with a four-lens turret, each valuable for a special purpose.

Essential Elements of the Television Camera

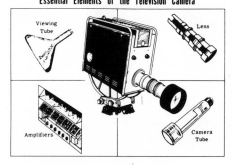

Viewing Tube · Lens · Amplifiers · Camera Tube

4. *Television Cameras* consist of four sections: lenses which pick up the image; the camera tube which converts light energy into electric energy; amplifiers; and the cameraman's viewing system.

5. *The Vidicon Camera* is one of the smallest and most versatile television cameras. Originally developed for outdoor and portable uses, it is now being adapted for regular studio purposes.

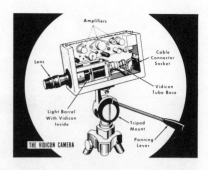

6. *The Television Studio* is unique for its vastness and complexity. Many sets are in operation simultaneously—one for a show on the air, others for shows in rehearsal.

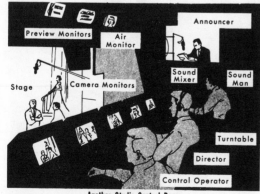

Another Studio Control Room

7. *The Director* carries the ultimate responsibility for the program. He works in the studio control room with control operators. Displayed in front of him is the output from each camera. He selects, via the master switching panel, the pickup he wants transmitted, or video-taped.

Tracing the Sound Signal

Let us now begin to examine how the sound and picture signals advance from the studio to the transmitting antenna. We shall deal first with the sound signal.

Whenever possible, microphones are placed out of sight; behind flowers, desk equipment, the folds of a dress, statues or any available prop. Stage directors have been very successful in this attempt to maintain illusion and prevent distraction, without sacrificing the quality of sound pickup. Even the color of the microphones is given consideration; those worn by narrators or interviewers who walk about the stage during a program, or those visible on a desk, are colored to be as inconspicuous as possible.

You may have wondered how the sound is picked up when the speaker is seen in full-length shot with no microphone cables visible. The trick is the use of a miniature short-distance microphone-transmitter. The microphone pickup modulates the tiny transistorized transmitter, which in turn radiates a signal picked up outside camera range.

Tracing the Sound Signal (cont'd)

Now we will construct, step by step, the sound signal block diagram from studio to transmitting antenna.

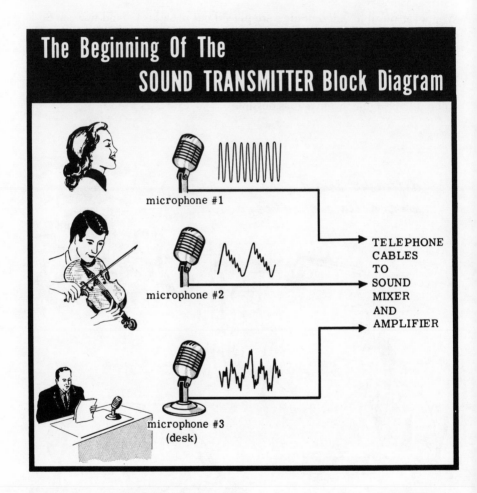

The microphones, waveforms and sources of sound indicated on the block diagram are only symbolic. The sound waves picked up by the microphones are translated into electrical voltages by these devices and are fed into the control room through individual audio cables. We show only three microphones but there may be more or fewer in use at any one time. There is no elevated boom-type microphone in this picture, although it is a frequently used device.

Adding the Sound-Mixer Amplifier

Microphones seldom pick up the required level of audio energy, and it becomes necessary to control the level of the microphone output. Provisions for raising or lowering the level to suit a particular situation are part of the audio amplification process—a function of the mixer-amplifier.

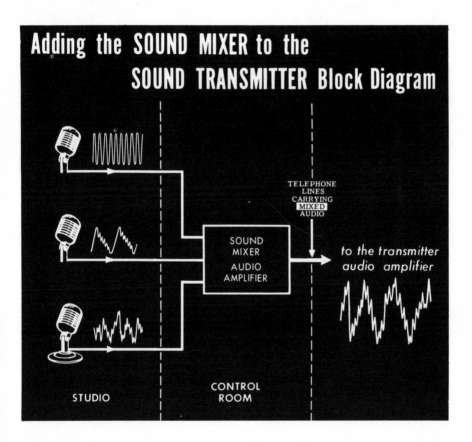

Still another function of the mixer is suggested by its name. Regardless of how many sources of sound are part of the televised scene, and how many microphones are used to pick up these sounds, it is a composite of *all* of them that becomes the audio voltage which modulates the carrier wave transmitted to the receiver. All the microphones feed into the mixer. There the input level of each microphone signal is individually controlled and fed into a common amplifier. The result is piped over telephone lines to the transmitter to be amplified more and applied to the FM carrier.

Producing the Frequency-Modulated Sound Carrier

To effectively broadcast television sound information, a suitable carrier frequency must be generated. The audio-frequency signals are then used to modulate (or change) the radio-frequency carrier. It is this modulated radio-frequency carrier that is then sent out from the TV broadcast antenna.

In standard radio broadcasting one method of combining the audio intelligence with the radiated carrier is *amplitude modulation*. The superimposition of the audio signal on the radio-frequency carrier results in a final carrier which is *constant in frequency* but *varies in amplitude* in accordance with the instantaneous variations of the audio-modulating voltage. Thus, the *envelope* of the amplitude-modulated carrier has the same outline as the instantaneous variations of the audio voltage.

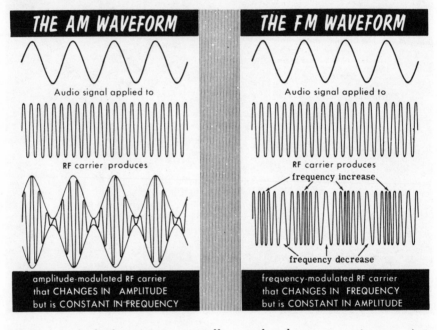

THE AM WAVEFORM | THE FM WAVEFORM

Audio signal applied to

RF carrier produces

RF carrier produces
frequency increase

frequency decrease

amplitude-modulated RF carrier
that CHANGES IN AMPLITUDE
but is CONSTANT IN FREQUENCY

frequency-modulated RF carrier
that CHANGES IN FREQUENCY
but is CONSTANT IN AMPLITUDE

Another method of transmitting intelligence by electromagnetic waves is to convert the audio voltage that is to be transmitted into frequency variations of the radio-frequency radiated carrier. This process delivers the sound portion of the television broadcast to the viewer. Such modulation is called frequency modulation and produces a radio-frequency carrier which is *constant in amplitude* but *varies in frequency* within certain maximum limits determined by the system design. The increase and decrease in frequency is a function of the *loudness of sound* (amplitude of the modulating voltage), whereas the *rate of change* of frequency is a function of the frequency of the *audio modulating voltage*.

Producing the Frequency-Modulated Sound Carrier (cont'd)

To understand the special language related to frequency modulation, consider the sound signal transmitted from TV Channel 5. For Channel 5 the RF frequency allocation is 81.75 MHz. Therefore, when the carrier is not being modulated by an audio-frequency signal, the sound carrier has a constant or "resting" frequency of 81.75 MHz.

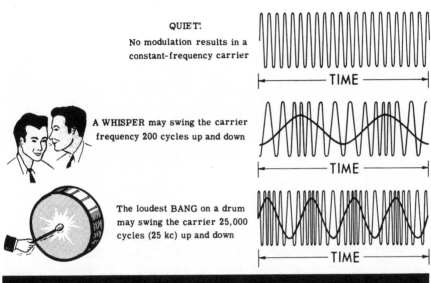

QUIET!

No modulation results in a
constant-frequency carrier

|—————— TIME ——————|

A WHISPER may swing the carrier
frequency 200 cycles up and down

|—————— TIME ——————|

The loudest BANG on a drum
may swing the carrier 25,000
cycles (25 kc) up and down

|—————— TIME ——————|

Frequency Modulation

The change in carrier frequency caused by the audio-frequency signal is called frequency *deviation. Positive deviation* occurs when the frequency rises above the resting frequency, while *negative deviation* occurs when the frequency drops below the resting frequency. The total deviation or the sum of the negative and positive frequency shifts is called frequency *swing.* Consider the following example: The slightest whisper into a Channel 5 microphone may cause a frequency shift of 200 Hz above and 200 Hz below the carrier frequency of 81.75 MHz. Thus, the whisper causes a total deviation or frequency *swing* of 400 Hz from 81.7498 to 81.7502 MHz.

On the other hand, a sound loud enough to produce 100% modulation of the FM carrier would, according to United States FCC standards, raise and lower the carrier frequency 25,000 Hz (25 kHz) for an overall frequency *swing* of 50 kHz. The 100% modulation limits for the Channel 5 resting frequency of 81.75 MHz would then be 81.725 MHz (81.750 − 0.025) for the lower limit, and 81.775 MHz (81.750 + 0.025) for the upper limit.

Producing the Frequency-Modulated Sound Carrier (cont'd)

In actual practice, the FCC requires that the sound carrier oscillator generate a relatively low-frequency signal. Frequency multipliers are then used to raise the carrier frequency to the desired frequency level. For example, to obtain the Channel 5 carrier frequency of 81.75 MHz, a sound carrier oscillator is used to generate a frequency of 2.5547 MHz. Then frequency multipliers are used to multiply the 2.5547-MHz frequency 32 times to produce the necessary 81.75-MHz carrier frequency.

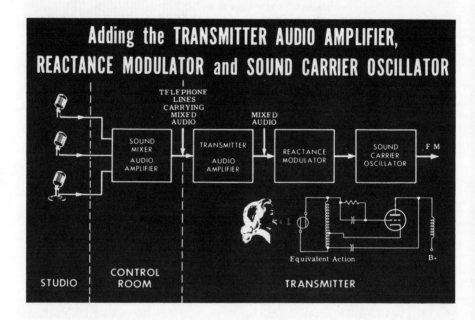

The reactance modulator and the sound carrier oscillator work as a unit to produce the frequency-modulated output. Audio signals from the transmitter audio amplifier feed into the reactance modulator which translates the audio voltage variations into frequency changes and impresses them on the sound carrier. It does this by acting as an automatically variable tuning capacitor that controls the sound carrier oscillator output. The variation in capacitance follows the audio-signal voltage changes and automatically varies the output frequency of the sound carrier oscillator. Thus the deviation in the carrier frequency corresponds to the *amplitude* variations in the audio signal, while the rate of change of the carrier-frequency deviation depends on the *frequency* of the audio signal. The amount of frequency swing allowed in the sound carrier oscillator during frequency modulation depends upon the frequency multiplication that will occur further along in the system. This is explained in the pages that follow.

Producing the Frequency-Modulated Sound Carrier (cont'd)

Frequency-Multipliers
The proper frequency for the radiated carrier is obtained by frequency-multiplication of the sound-carrier oscillator frequency.

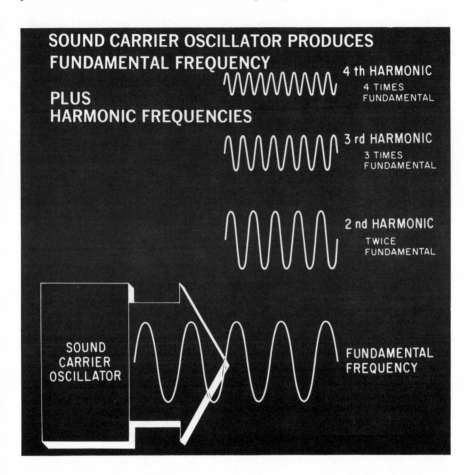

A conventional vacuum-tube oscillator generates not only the fundamental frequency voltage for which it is designed but also a host of other voltages at frequencies known as *harmonics* or *harmonic frequencies*. Harmonics are simply voltages whose frequencies are *whole number multiples* of the fundamental frequency. Assume for the moment a fundamental frequency (f) of 2.5 MHz. The second harmonic or $2 \times f$ will have a frequency of 2×2.5 or 5 MHz; the third harmonic or $3 \times f$ will have a frequency of 3×2.5 or 7.5 MHz; the fourth harmonic or $4 \times f$ will have a frequency of 4×2.5 or 10 MHz, etc. Depending on the equipment, harmonics up to the 200th may be useful.

Producing the Frequency-Modulated Sound Carrier (cont'd)

The fundamental-frequency oscillator is followed by an amplifier which accepts all the frequencies present in the oscillator output. But because it contains a tuned circuit in its plate circuit, resonant to the second harmonic of the fundamental, the amplification is effective only on the second harmonic. Such an amplifier functions as a frequency-doubler. Its output signal frequency is equal to twice the frequency fed to the input.

Amplifiers... when used

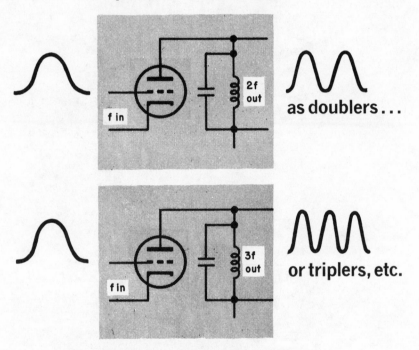

as doublers...

or triplers, etc.

are FREQUENCY MULTIPLIERS

In a similar manner an amplifier may be arranged to amplify the third harmonic of an input frequency, in which case it is acting as a frequency-tripler, or it may select the fourth harmonic for amplification and then is a frequency-quadrupler. When referred to in broad terms such amplifiers are frequency-multipliers. Two, three, four or more frequency-multipliers may follow each other, thus affording an overall frequency-multiplication of whatever amount is needed. Each amplifier is arranged to accept whatever frequency is fed into it and to distort the signal, thus creating harmonics within the amplifier stage. The tuned plate circuit then selects the harmonic desired for transfer to the next frequency multiplier.

Producing the Frequency-Modulated Sound Carrier (cont'd)

A series of frequency-multipliers raises the frequency-modulated sound-carrier oscillator output signal to the desired value (as determined by the FCC allocation of the sound-carrier frequency) to a television station. Assuming channel 5 as the station being discussed, the allocated sound-carrier frequency of 81.75 MHz can be achieved by generating a 2.5547-MHz fundamental frequency in the sound-oscillator and multiplying this frequency 32 times; 2.5547 × 32 = 81.7504 MHz, in round numbers 81.75 MHz.

The 32-time multiplication can be accomplished in several ways: in three steps such as 2 × 4 × 4 using a doubler and two quadruplers. The signal outputs would have frequencies of 2 × 2.5547 or 5.1094 MHz, 4 × 5.1094 or 20.4376 MHz and finally 4 × 20.4376 or 81.7504 MHz. Or four frequency-multiplier stages such as 2 × 4 × 2 × 2, or 2 × 2 × 2 × 4 could be used.

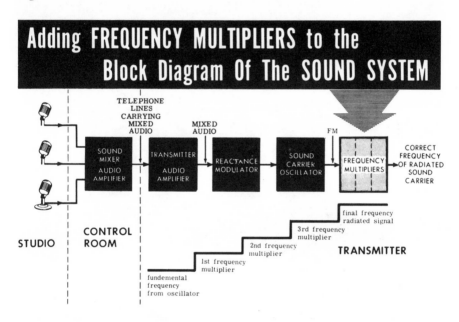

There is one very significant detail that relates the amount of frequency-multiplication applied to the fundamental sound-carrier frequency to the change in frequency corresponding to the frequency-modulated radiated sound carrier. Inasmuch as the maximum change in frequency of the radiated sound carrier is 25 kHz in each direction for 100% modulation, 32-time multiplication of the fundamental frequency limits the frequency changing action by the reactance-modulator on the sound-carrier oscillator signal of 2.5547 MHz to 1/32 of 25 kHz or 781 Hz. If the amount of frequency-multiplication is different, the frequency change corresponding to 100% modulation of the fundamental sound-carrier oscillator is modified.

The FM Sound Power Amplifier

There is just one more amplifier stage that the frequency-modulated sound carrier must pass through before it is "put on the air." This is the power amplifier. To make certain that the radiated signal has ample strength when it reaches the receiver, the amount of electrical energy fed to the transmitting antenna must be of a prescribed level. Achieving this is the function of the tuned power amplifier stage, which makes use of high-power tubes. Depending on the power output rating of the transmitting station, these power amplifiers may be rated from a few thousand watts to perhaps 25 to 50 kW.

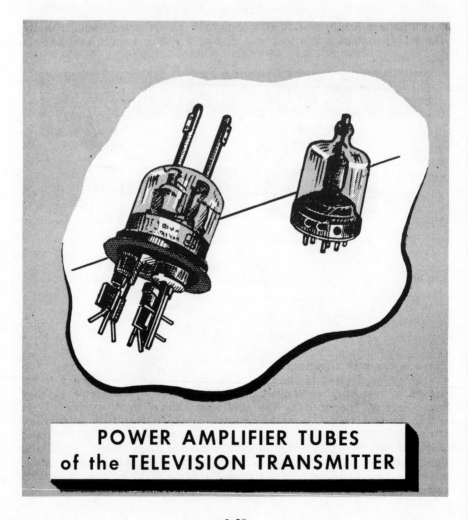

POWER AMPLIFIER TUBES of the TELEVISION TRANSMITTER

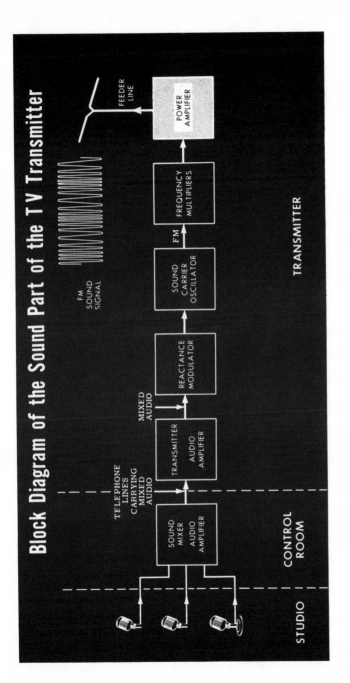

Block Diagram of the Sound Part of the TV Transmitter

Complete Sound System in TV Transmitter

This block diagram of the sound system in the television transmitter is more of a symbolization than an attempt to show exactly the amplifier stages. Some transmitters use a single antenna for radiating the sound and the picture carriers. Others use separate antennas for putting the two signals on the air. The feed line between the power amplifier and the antenna is an RF transmission line capable of carrying the required amount of RF energy. Details about transmission lines are given later in the course.

The Picture System

The start of the picture system in the television transmitter is the camera. The heart of the camera is the camera tube. As our first step in the explanation of this tube we shall discuss the most fundamental type of tube which can convert light energy into electrical energy. This is the phototube.

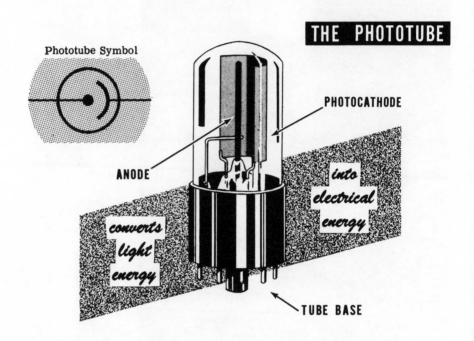

THE PHOTOTUBE

Phototube Symbol

PHOTOCATHODE

ANODE

into electrical energy

converts light energy

TUBE BASE

The phototube looks like a conventional radio receiving type of vacuum tube. The semicircular electrode inside the envelope is the photocathode. The upright metal electrode near the curve of the photocathode is the anode. The schematic representation of the tube bears a close resemblance to its physical structure as would be seen from a top view. Electrical connections to the electrodes is by means of conventional tube pins protruding from the tube base—in all respects like the conventional receiving tube base.

The derivation of the word photocathode is the following: "photo" from the Greek meaning light, and "cathode," the name for an emitter of electrons. A photocathode then is simply an electrode bearing a surface or coating which has the property of emitting electrons when struck by light. The anode, on the other hand, is a name assigned to an electrode that attracts electrons to itself by a positive voltage applied to it. In this case it attracts the electrons that are emitted by the photocathode. Thus, a *phototube is a device which converts light energy into electrical energy.*

Phototube Circuitry and Functioning

The basic phototube circuit consists of the tube, a load resistor and a source of dc voltage. A source of light is of course required but it is not a part of the phototube circuit. When light shines on the inner surface of the photocathode the coating emits electrons. The positive dc voltage derived from the power supply and applied to the anode causes the anode to attract the emitted electrons which flow through the circuit in the direction shown by the arrow. The brighter the light that shines on the phototube the higher is the amount of current flowing in the circuit. In the absence of any light the phototube current is zero. Since the tube current flows through the load resistor (R) a voltage drop proportional to the tube current is developed across the load. The output voltage is the electrical energy equivalent of the light energy that strikes the photocathode. Hence, the tube is a converter of radiant (light) energy into electrical energy.

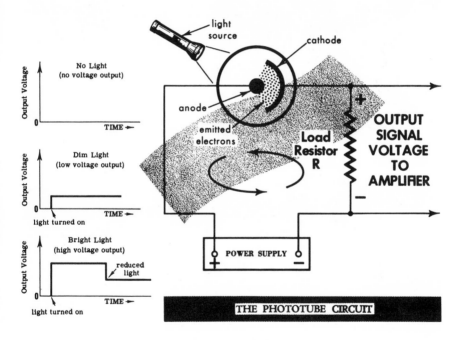

THE PHOTOTUBE CIRCUIT

The output voltage graphs symbolize the behavior of the phototube under different conditions of incident light. It is seen that the output voltage is zero when incident light is zero. If the light is dim the output voltage is low; if the light is bright the output voltage is high. A sudden change in light intensity from bright to a lower level results in a corresponding instantaneous change in output voltage. Thus the phototube can produce an output signal voltage which instantaneously increases and decreases in amplitude in accordance with the changes in intensity that strike the photocathode.

Primitive Image-to-Voltage Conversion System

A primitive system for converting the image of an object into video signal voltages is discussed here. ("Video" is from the Latin, "I see.") It is an impractical system, but useful to us as a foundation for understanding the action in the image pickup tube of a television camera.

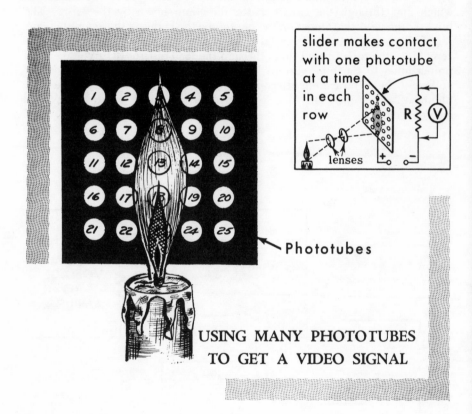

slider makes contact
with one phototube
at a time
in each
row

lenses

Phototubes

USING MANY PHOTOTUBES
TO GET A VIDEO SIGNAL

Imagine 25 small phototubes assembled on a board located in back of a lens system. All the anodes are connected to the power supply, but the photocathodes (abridged to cathodes) end in separate terminals so that each tube cathode can be contacted individually by a mechanically driven slider. The slider moves along each horizontal row of cathode terminals, contacting one at a time and completing a tube current path. Having completed the top horizontal row it automatically and very rapidly positions to start the next lower row, and repeats the horizontal scan. This action is repeated for each horizontal row, one row at a time. A recording voltmeter (V) indicates the waveform of the voltage drop developed across the load resistor(R).

Primitive Image-to-Voltage System

The image that is projected onto our bank of phototubes is that of the flame of a candle. For identification in the discussion we have assigned arbitrary numbers to each of the phototubes. It is seen that the image of the flame covers some of the phototubes completely, others are unaffected, and still others bear only a portion of the image of the flame. This, plus the difference in brightness of the core of the flame, means that light varying in amount from zero to maximum impinges on the phototubes. Each phototube will produce a dc voltage proportional to the light upon it.

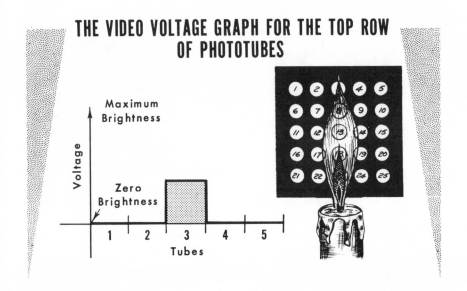

THE VIDEO VOLTAGE GRAPH FOR THE TOP ROW OF PHOTOTUBES

All portions of even the bright parts of the flame are not equally bright, but this elementary system of using a few phototubes cannot resolve segments of parts of the flame. Therefore the output video voltage from each phototube represents the average amount of illumination incident on the photocathode. This limitation in performance does not hinder our examination, for all that interests us at the moment is the development of the output video voltage, horizontal line by horizontal line. The video voltage graph for the top row shows zero voltage output for tubes 1, 2, 4, and 5, and an arbitrary value (less than maximum) for tube 3, which is subject to illumination from only the tip of the flame. The abrupt change of voltage from zero illumination (tube 2) to partial illumination (tube 3) and return to zero illumination (tube 4) results in a straight-line rise and fall of the output video voltage. The interval between phototubes is neglected in this voltage graph.

Primitive Image-to-Voltage System (cont'd)

We scan each horizontal row of phototubes and develop a series of output video voltages, line by line. The amount of illumination incident on tube 8 (line 2) is greater than tube 3 (line 1). Therefore, the amplitude of the output voltage developed by the current from tube 8 is greater.

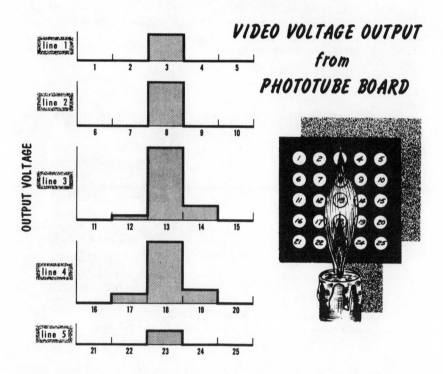

VIDEO VOLTAGE OUTPUT from PHOTOTUBE BOARD

Looking at line 3, tube 12 is subject to just the fringe of the flame, therefore its output voltage is low. Tube 13 is brightly illuminated, therefore the output voltage is maximum. Tube 14 is illuminated less than tube 13 but more brightly than tube 12, so its output voltage is appreciably less than tube 13, but more than tube 12. In the fourth line tubes 17, 18, and 19 are illuminated, but to different amounts, hence the output voltages have different amplitudes. The illumination on tube 18 is partly from the core of the flame, hence the output voltage is less than the maximum, as for example from tube 13. On the other hand tube 17 is subject to more illumination than 12, so its voltage output is greater. The reverse is true of tubes 14 and 19.

Finally we scan line 5. Here only tube 23 is illuminated, and that not brightly, because it is the core of the flame. The output voltage is relatively low. It should be understood that the levels of the video voltages shown are strictly arbitrary and entirely illustrative.

Primitive Image-to-Voltage System (cont'd)

Let us now elaborate on the primitive system we have shown by increasing the number of phototubes onto which the image is projected. What we shall end up with is not a duplicate of the commercial televison camera tube systems, but it will approach the system closely enough to render the entire operation more understandable.

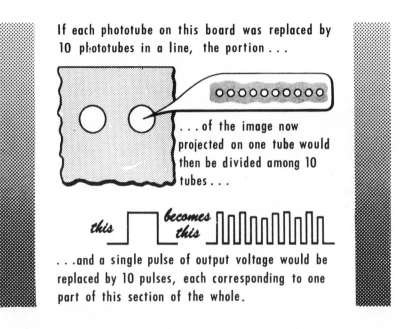

If each phototube on this board was replaced by 10 phototubes in a line, the portion . . .

. . . of the image now projected on one tube would then be divided among 10 tubes . . .

this *becomes* *this*

. . .and a single pulse of output voltage would be replaced by 10 pulses, each corresponding to one part of this section of the whole.

If, in the space previously occupied by one phototube we place 10 phototubes which are each separate from the other, the portion of the candle flame image previously projected onto one phototube now would illuminate 10 phototubes. The illumination on each of these tubes would be representative of a section of that part of the whole flame. Each tube then would produce an output voltage corresponding to a much smaller part of the whole image than when fewer but larger phototubes were used. Differences in brightness of parts of the flame would create voltage pulses of different amplitude, whereas when all of these 10 parts illuminate just one phototube the voltage pulse represents an average of the total illumination. In other words, the greater the number of parts onto which an image is divided the greater is the detail which is possible in the reproduction of the image because each part then produces its own voltage pulse. As will become evident the commercial version of the photocell pickup device as used in the television camera tube divides the image into a great many small elements, many thousands of times more numerous than shown even in this elaboration of the primitive system.

Zworykin's Iconoscope

Vladimir Kosma Zworykin, then Director of Research at the Radio Corporation of America, developed the first practical television camera in 1938, thus converting the primitive phototube board into a functional device. Inside of an evacuated glass envelope, Zworykin placed an assembly of components which formed a successful television camera tube. Although modern camera tubes do not even remotely resemble the original iconoscope even in principle, we discuss it in detail here because it will help you make the transition from one camera tube type to the next with greater facility. Zworykin gave the tube the name *iconoscope*, derived from "icon" meaning image and "scope," to observe. He called the phototube board the *mosaic*.

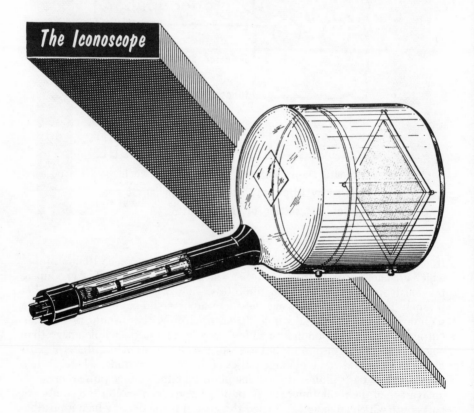

The Iconoscope

The Mosaic

Strange names such as mosaic are chosen in electronics because of their descriptive meanings. In the world of art, mosaic is the name for a pattern formed by inlaying small pieces of colored stone, glass, or other material.

The Iconoscope (cont'd)

The mosaic in the iconoscope actually consists of millions of tiny drops of photosensitive compound deposited on one side of a thin sheet of mica. Each photodroplet is capable of emitting electrons when struck by light, so in effect we have millions of tiny phototubes arranged on the mica sheet. On the other side of the mica sheet is a thin coat of conducting graphite called the signal plate. The mosaic board resembles a sandwich, with light-sensitive cells on one side, the signal plate on the other side, and nonconducting mica in between. In every sense this arrangement is equivalent to a capacitor. In fact each photodroplet and the corresponding area of the signal plate does comprise a tiny capacitor. Thus the mosaic board is made up of millions of tiny photo-emissive capacitors with one active surface of each capacitor, the signal plate, common to them all.

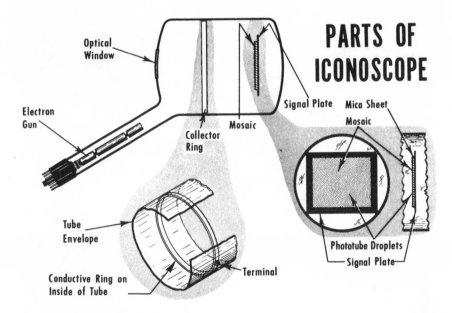

The Collector Ring

You will recall that every phototube has an electron-collecting electrode, the anode. A similar electron-collecting device common to all the photodroplets is part of the iconoscope, and called the collector ring. It is a metallic coating in the form of a narrow ring on the inside of the iconoscope envelope and is located a short distance from the mosaic. Each photodroplet, together with the collector ring, forms a tiny phototube in which the droplet is the photocathode and the collector ring is the anode. Zworykin's mosaic, with its millions of phototubes, is the professional version of the primitive phototube board. With so many picture element pickups, it is capable of developing a highly detailed picture.

The Iconoscope Circuit

In every way the iconoscope is more elaborate than a simple phototube, and its electrical system is far more complex. So that we may easily understand its functioning we shall divide the action of the iconoscope into two parts: the illumination of the mosaic by the scene, and the electron-beam scanning-action. The former is related to the storage of the image on the mosaic, the latter is related to the conversion of stored picture information into picture voltage pulses. Associated with these actions is the behavior of each photodroplet and the signal plate as a tiny capacitor.

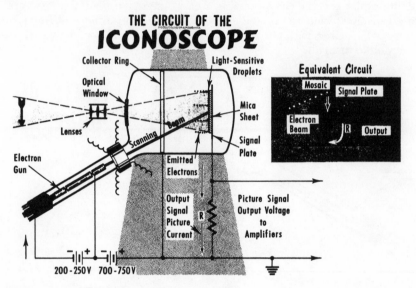

Assume that an image is projected onto the photodroplets. Each tiny phototube will emit electrons in proportion to the amount of light incident on it. The emitted electrons are attracted to the collector ring. At the same time there occurs a redistribution of electrons through the external circuit between each droplet and its corresponding facing area on the signal plate. In other words, each tiny capacitor becomes *charged*. The charge corresponds to storage of that portion (element) of the picture illumination which was incident in the droplet.

The electron gun in the iconoscope produces a very narrow pencil of electrons. It is aimed at the mosaic and by suitable means (deflection, to be explained later) the electron beam moves (scans) horizontally across the photodroplets, describing one very thin horizontal line at a time. The scanning electron beam replenishes the electrons lost by each photodroplet, that is, it suddenly discharges each photoemissive capacitor. The discharge current is the picture current that flows through the load resistor (R) and develops the picture signal output voltage applied to the amplifiers.

The Electron Gun

The electron gun is aptly described by its name; it is an assembly of components designed to create a thin, dense beam of electrons and to propel them at high velocity to the target. In this instance the target is the mosaic. Heated to a high temperature by an associated incandescent heater, the cathode emits electrons in abundant supply. The quantity of electrons which makes up the beam is determined by the control grid, an electrode lying in the path of the emitted electrons and located near the cathode. Having passed the control grid the electrons enter two cylindrically shaped electrodes, anode 1 and anode 2.

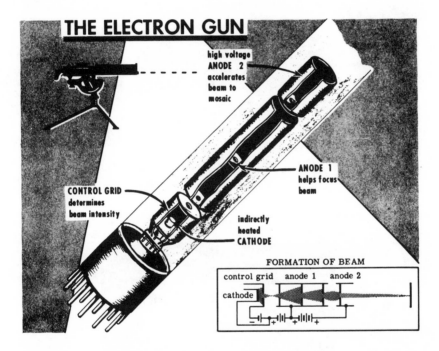

Both anodes are subject to dc voltages, positive with respect to the cathode. They exert an attracting force on the electrons and propel them towards the mosaic. The voltage applied to anode 1 is much less than that applied to anode 2. Because of the shapes of these electrodes and the voltages applied, the electrons moving from anode 1 towards anode 2 become densely packed into a thin beam. The action is called focusing. It results in an electron beam that has a very narrow diameter (about .015 in.) at the point where it strikes the mosaic. In addition to aiding the focus, the high positive voltage applied to anode 2 causes electrons to advance towards the mosaic at a high speed. As indicated in the illustration, the complete electron gun assembly is located inside the neck of the tube.

The Deflection Yoke for the Iconoscope

The electron beam in the iconoscope releases picture information stored in the mosaic's photodroplet capacitors. It moves across the entire surface of the mosaic line by line, eventually discharging all the photodroplets. The beam motion is *to and fro horizontally* and *depressed vertically*. Each horizontal excursion and return is lower on the mosaic than the preceding one. Finally the electron beam reaches the bottom, and from there is returned to the top to start the next series of horizontal movements. Such motion of the electron beam across the surface of the mosaic is called *scanning the mosaic*. When the beam performs these movements it is said to be *deflected*.

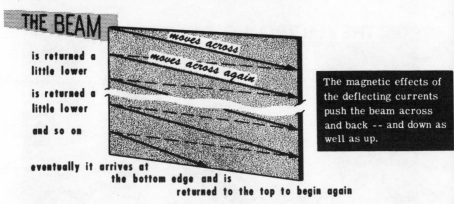

THE BEAM

is returned a little lower

is returned a little lower

and so on

moves across

moves across again

eventually it arrives at the bottom edge and is returned to the top to begin again

The magnetic effects of the deflecting currents push the beam across and back -- and down as well as up.

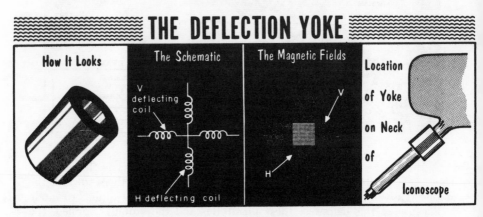

THE DEFLECTION YOKE

How It Looks	The Schematic	The Magnetic Fields	Location
	V deflecting coil	V	of Yoke
	H deflecting coil	H	on Neck of Iconoscope

The forces which move the beam horizontally and vertically, (the deflecting forces), are produced by two externally produced magnetic fields arranged to penetrate the neck of the iconoscope and so interact with the magnetic lines of force that encircle the tube's electron beam. The two external magnetic fields originate in the horizontal and vertical deflecting coils which make up the *deflection yoke*. Each coil carries suitable deflection currents. The yoke as a whole is positioned around the neck of the iconoscope.

1. *The Audio System* originates with the microphone. Different types of sound (music, voice, sound effects) necessitate different microphones. After picking up sound waves and translating them into electrical voltages, the mike feeds the audio through a cable into the control room.

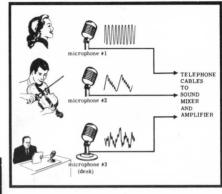

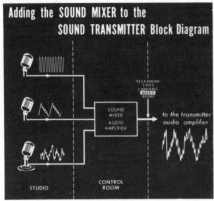

2. *Amplification.* The mixer-amplifier in the studio receives, then amplifies, the sound signal. It also blends several sound sources into one signal. From the mixer, the sound is piped through telephone lines into the transmitter.

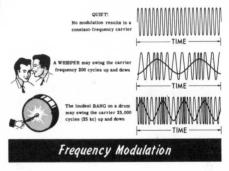

3. *Frequency Modulation.* Television audio intelligence is transmitted by FM. With this method, audio voltages are converted into frequency variations of the RF radiated carrier, which is constant in amplitude but variable in frequency.

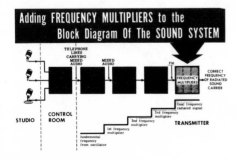

4. *Frequency Multiplication.* The sound-carrier oscillator generates both fundamental and harmonic frequencies. They are sent to a series of amplifiers, each of which is effective on only the harmonic fed to it. A tuned plate circuit selects the harmonic to be transferred to the next frequency multiplier.

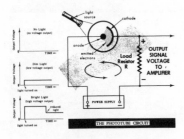

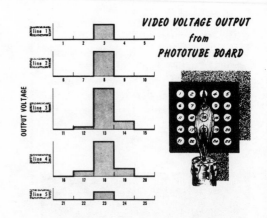

VIDEO VOLTAGE OUTPUT *from* PHOTOTUBE BOARD

5. *Phototube Circuit.* The heart of the camera is the camera tube. The most basic type of camera tube is the phototube. When light shines on the photocathode, electrons are emitted which are attracted by the anode. The strength of the resultant flow varies with the strength of the original light.

6. *On a Phototube Board* each tube is exposed to one section of the image. A slider is mechanically driven across the terminal of each tube, developing a series of video voltages.

7. *The Iconoscope* uses a *mosaic*—a sheet of mica covered with a photosensitive compound. In the iconoscope, the electron-attracting anode takes the form of a collector ring, a metallic coating inside the iconoscope envelope neck.

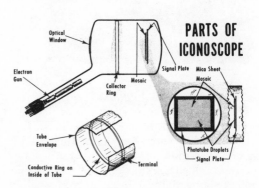

PARTS OF ICONOSCOPE

8. *The Electron Gun* directs a beam of electrons, emitted by the cathode, at the mosaic. The electrons pass two attracting anodes of unequal voltages which propel the electrons towards their target, and also cause them to become densely packed, aiding in focus.

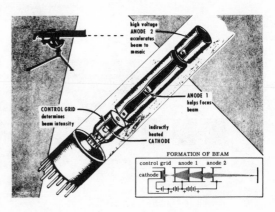

The Image Orthicon

In many television station studios, iconoscopes are being replaced by another
Zworykin development called the image orthicon. One important reason for
the decline in popularity of the iconoscope is its lack of sensitivity and the con-
sequent need for intense illumination of the scene. A second, more important
reason is its tendency to generate false signals.

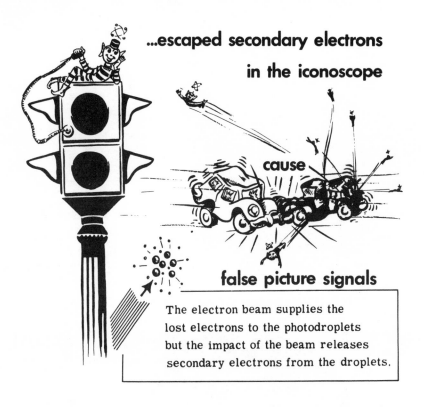

...escaped secondary electrons

in the iconoscope

cause

false picture signals

The electron beam supplies the
lost electrons to the photodroplets
but the impact of the beam releases
secondary electrons from the droplets.

Secondary Emission
The false-signal phenomenon is caused by secondary emission of electrons
from the photodroplets. The scanning beam supplies electrons to the
photodroplets of the mosaic and so releases the picture information, but the
impact of the beam when it strikes the droplets also has another effect. It
causes the photosensitive material to emit secondary electrons. They scatter in
all directions and fall on neighboring droplets which have not yet been
scanned. The result is a charge which is not caused by the image. Thus the
total charge on these droplets is not true picture information; when these
droplets are discharged by the scanning beam, the result is current due to the
image charge plus secondary electron charge.

The Image Orthicon (cont'd)

Secondary emission that occurs when the scanning beam strikes the iconoscope mosaic is certainly unwanted. Yet there are occasions when secondary emission can be put to good use. In the image orthicon (the camera tube which succeeded the iconoscope) Zworykin used a low-speed scanning beam to avoid the creation of secondary emission. In another part of the tube he used an electron multiplier, an assembly of tube electrodes, which is useful *because* of the secondary emission of electrons. It utilizes secondary emission to amplify the picture information current and thus the image orthicon is a more useful and valuable camera tube than is the iconoscope.

With like amounts of illumination from an image, the image orthicon affords a much greater picture current output than does the iconoscope. A sort of chain reaction is created wherein the secondary electrons that are emitted are captured and put to work to cause greater current output than input. This action, repeated several times, occurs in the electron multiplier. In addition to its high current output, the image orthicon is free of false signal output.

The Image Orthicon (cont'd)

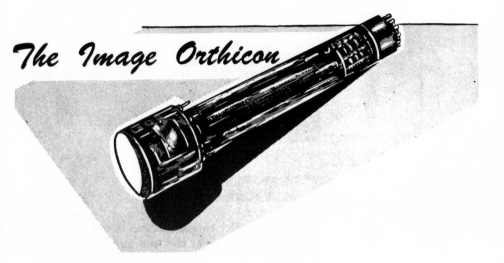

The image orthicon differs from the iconoscope in appearance and mode of operation, but there are some fundamental similarities between the two. In the image orthicon the neck of the tube containing the electron gun lies along the long axis of the tube, with the electron gun at one end. On the other end is located the optical window through which the rays focused by the lens system are admitted to the tube. Between the two is located the target. The image is produced on one side of the target in a special manner soon to be described, and the scanning beam originating in the electron gun strikes the other side of the target.

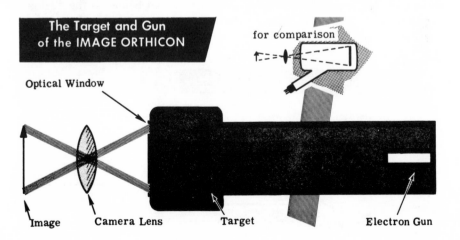

Adding the Photocathode

The image orthicon *mosaic* is really two separate electrodes: the *target* that is scanned by the beam, and the *photocathode* which emits electrons when struck by light. The photocathode is a translucent surface located inside the tube envelope near and parallel to the optical window. The lens takes the light reflected from the object being televised and focuses it, as a sharply defined image, onto the photocathode. As the image falls on the front side of the photocathode the rear surface emits electrons in quantities proportional to the light striking the point. Because the photocathode is negative relative to the target, the electrons advance towards the target.

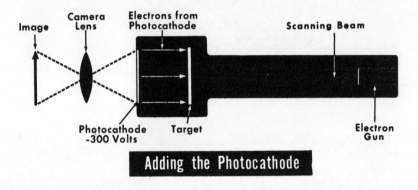

Adding the Photocathode

Adding the Focus Coil

The electrons emitted from the photocathode advance towards the target. Proper functioning of this camera tube requires that the electrons move in straight parallel lines. To accomplish this, the tube uses a dc current-carrying focus coil. The magnetic field created by the current bends all the diverging electrons into the correct paths so they advance between the photocathode and the target in parallel paths.

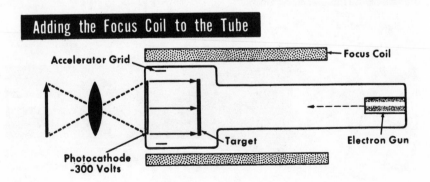

Adding the Focus Coil to the Tube

Behavior of the Target

The target is a rectangular sheet of very thin glass, specially composed. Its *surface* resistance is very high in comparison to the resistance *through* the glass from one side to the other. The difference in conductivity along the surface and through the sheet is of great importance.

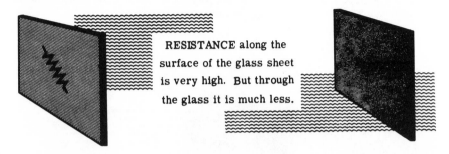

RESISTANCE along the surface of the glass sheet is very high. But through the glass it is much less.

If a beam of electrons strikes a point on the surface of the glass it dislodges secondary electrons from that tiny area, causing a positive charge. Because of the high surface resistance the charged condition remains intact and will neither dissipate nor spread to adjacent areas along the surface. Thus if many individual rays of electrons strike the surface at the same time, there will be created many tiny charged areas, each isolated from the other.

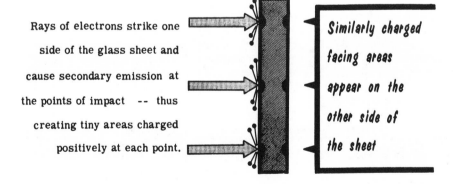

Rays of electrons strike one side of the glass sheet and cause secondary emission at the points of impact -- thus creating tiny areas charged positively at each point.

Similarly charged facing areas appear on the other side of the sheet

At the same time the thinness of the sheet and the conductivity through the glass causes the appearance of an equal number of similarly charged areas to appear on the other side of the sheet. Each charged area on one side of the sheet faces a charged area on the opposite side of the sheet. The magnitude of the positive charge created at each charged area is a function of the intensity of the electron ray that strikes the glass at each point. The greater the number of electrons in the bombarding ray, the greater the amount of positive charge created at corresponding points on each side.

Behavior of the Target (cont'd)

The location of the image orthicon target causes it to be bombarded by rays of electrons emitted by the photocathode. The density of each of these rays is determined by the amount of illumination incident on the photocathode at each emitting point. By making the emitted electrons travel in parallel paths after they have left the photocathode, the advancing rays form an electron image of the pattern which is on the photocathode.

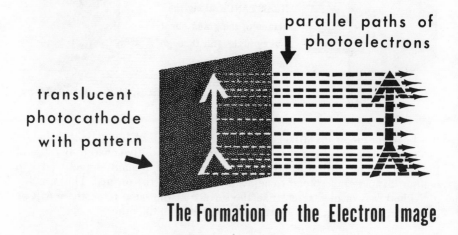

parallel paths of photoelectrons

translucent photocathode with pattern

The Formation of the Electron Image

When the electron rays strike the target they form a charge image of whatever pattern is on the photocathode on *both* sides of the target. As we stated earlier the electrons strike the target on the photocathode side only, but the conductivity of the glass results in the reproduction of charge images on the other side of the target. So in a sense the target in the image orthicon is a two-sided mosaic. The reason for this is the removal of picture information by scanning of the mosaic with an electron beam.

The Formation of the Charge Pattern on the Target

TRANSLUCENT PHOTOCATHODE WITH IMAGE OF WHITE ARROW

ELECTRON IMAGE

TARGET

CHARGE PATTERN OF IMAGE APPEARS ON BOTH SIDES OF TARGET

The + Signs Indicate Positive Charge

The Target Screen

The secondary electrons emitted from the target must be captured. This is done by a screen of very thin wire located near the target. A low voltage of positive polarity is applied to the screen. While the screen is effective in catching the secondary electrons from the target, it does not block the passage of electron rays coming to the target from the photocathode.

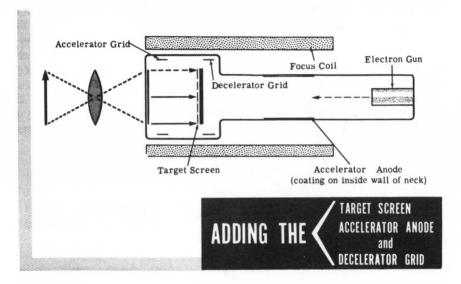

The Accelerator Anode

The iconoscope utilizes a scanning beam that is formed from the electrons emitted by the electron gun. To make the beam move to the target an accelerator anode is included as part of the tube assembly. It is a conductive coating of graphite on the inside wall of the tube neck. A relatively high positive voltage applied to this anode attracts the electrons and accelerates the beam as a whole towards the target.

The Decelerator Grid

Acting under the influence of the high positive voltage applied to the accelerator anode, and without any form of control, the electrons in the scanning beam would strike the target with impact sufficient to free secondary electrons and so defeat the aims of the tube design. To prevent this from occurring a decelerator grid (actually a coating on the inside wall of the tube) is located near the target. A relatively low positive voltage is applied to it. The electron beam leaving the high-intensity electrostatic field of the accelerating anode and entering the low-intensity electrostatic field of the decelerator grid is greatly reduced in acceleration. Thus it strikes the target with sufficiently low impact for the secondary electrons not to be emitted.

The Behavior of the Scanning Beam

Although the scanning beam in the image orthicon is produced by an electron gun similar to that used in the iconoscope, and it is moved back and forth across the target by a deflection yoke placed around the neck of the tube, its overall action is quite different from that of the iconoscope.

When the electrons emitted by the gun enter the neighborhood of the accelerating anode they are immediately speeded up. Coming into the field of the deflection yoke, the beam feels a pull (a deflecting force) due to the magnetic fields created by the horizontal and vertical deflecting coils inside the yoke, and it is bent out of its original path towards the target. When the beam enters the area of the decelerating grid it slows down under the influence of the lower-level electrostatic field in the vicinity of this grid, but it still continues on the way to the target. On approaching the target, of zero voltage relative to the cathode, the beam slows up even more.

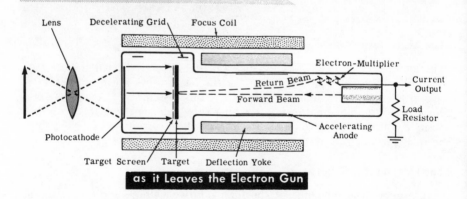

The Path of the Image Orthicon Scanning Beam

as it Leaves the Electron Gun

At this point let us divide the rest of the action into two parts. First we shall say simply that the beam strikes the target and in effect bounces off the target. Feeling the pull of the decelerating grid which is at a positive voltage, the beam reverses its direction and moves back towards its point of origin, but not along the same path. Passing the area of the decelerating grid, it feels the urge of the positive voltage applied to the accelerating anode and speeds up greatly, arriving at the collecting electrode. About this we say more later.

The Behavior of the Scanning Beam (cont'd)

Let us now consider the second part of the action. The beam has arrived at the target at one particular tiny area (A) where a positive charge corresponding to the image exists. The beam gives up as many electrons as are required to neutralize the positive charge at the point of impact. The remaining beam electrons which struck the target leave and move back in the general direction of the electron gun, but not quite *to* it. This beam now has in it fewer electrons than it had before it struck the target. So at this instant the returning beam is less dense than it was when it began.

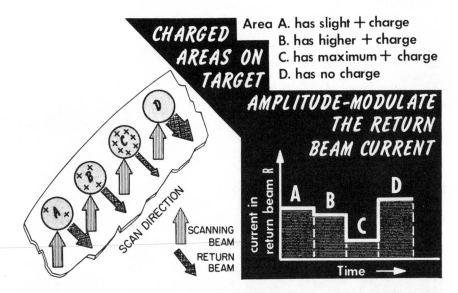

CHARGED AREAS ON TARGET

Area A. has slight + charge
B. has higher + charge
C. has maximum + charge
D. has no charge

AMPLITUDE-MODULATE THE RETURN BEAM CURRENT

SCAN DIRECTION

SCANNING BEAM

RETURN BEAM

current in return beam R

A B C D

Time →

An instant later the beam arriving from the electron gun strikes another tiny area (B) on the target, a point where there is now a greater positive charge. The beam gives up more electrons than before because it requires more electrons to neutralize the greater positive charge. Hence, upon return, the beam contains fewer electrons. Considering only these two instances, and recognizing that the electron beam is a current flowing through space, it is clear that the two return beams represent different amounts of picture current which correspond to the picture content appearing as a *charge image* at the two areas on the target struck by the beam.

If we now visualize the beam moving across the target and striking many points on the target, we can see that the return electron beam will momentarily increase and decrease in density because of the changing amount of electrons it gave up to neutralize the different areas on the target. In effect it will be amplitude modulated by the differences in the charge pattern on the mosaic, which correspond to the light and dark areas of the image.

The Electron Multiplier in the Image Orthicon

We have said that one major advantage of the image orthicon over the iconoscope is its greater sensitivity, accounted for by the electron-multiplier portion of the tube assembly. This device is a current amplifier.

The current amplifier consists of a number of metallic surfaces (dynodes) each bearing an increasingly higher positive voltage, and each emitting secondary electrons in abundance when struck by primary electrons. The physical organization of the plates is such that a primary beam (the return beam current) striking one plate causes the emission of secondary electrons which advance to the next plate. There they cause secondary electrons to be emitted in greater quantity, etc. This action is repeated among a number of plates, until the final stream of electrons delivered to a collector plate constitutes a much higher current than the original primary beam current.

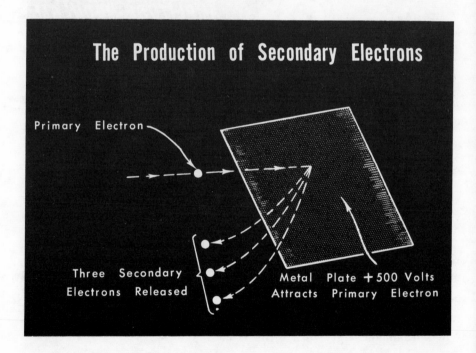

For an illustration, imagine a single electron traveling at high velocity towards a small metal plate. On impact with the metal plate the single electron knocks out three secondary electrons. If we can capture these three electrons, the current they constitute will be three times the primary current, and we will have a simple example of current amplification.

The Electron Multiplier in the Image Orthicon (cont'd)

Now suppose that we put a second metal plate near the first one and apply a higher positive voltage to the second plate. The three secondary electrons knocked out of the first plate will move at high velocity to the second plate, where each one of them will in turn liberate three new electrons from the surface. Now we have nine electrons in place of three, or if we think back on the single primary electron we have multiplied the current nine times.

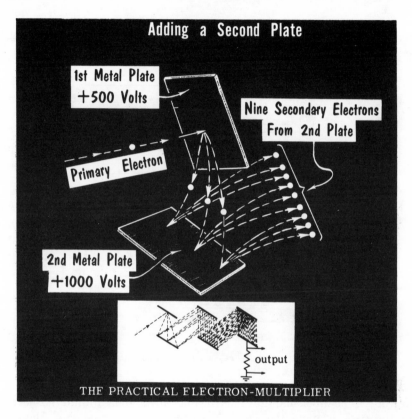

The commercial version of the electron multiplier may have as many as five stages (stages meaning electrodes which emit secondary electrons). The technical name for each of these metallic emitters is dynode. Current amplification may be achieved as many as 500 times. The last electrode in the assembly is the collector plate which accepts the abundant supply of electrons from the last dynode and passes it out of the tube as picture signal current. It should be understood that whatever variations exist in the primary stream of electrons that strike the first dynode will be maintained in the beam current returning from the target. These variations are repeated in the action which occurs at each dynode, and appear at the collector plate.

The Vidicon Camera Tube

The vidicon, a widely used camera tube, is substantially smaller than the image orthicon. It is highly adaptable to portable field and industrial applications in which the resolution requirements are less severe. This tube requires a scene illumination about eight times that of the image orthicon, but is very popular where reduced resolution and greater brightness of illumination are not serious handicaps.

The *Vidicon Tube* and its Structure

The front of the tube comprises three distinct layers going toward the electron gun: the glass faceplate, the signal electrode (which is a conductive film overlaying the inner surface of the faceplate), and a photoconductive layer deposited on the signal electrode.

The photoconductive layer consists of a mosiac of elements which are nonconductive in the dark but become electrically conductive when illuminated. The extent of their conductivity depends upon the intensity of illumination. As the image of the scene falls on the faceplate, the illuminated elements become more conductive and experience a change of voltage, coming closer to the charge on the signal plate. Since the signal plate is positive, a charge pattern develops on the signal plate which electrically duplicates the light image.

The Vidicon Camera (cont'd)

The beam from the gun is focused on the photoconductive layer by the long focusing electrode (grid 3) and the wire screen (grid 4) immediately in front of the inner surface of the faceplate. Scanning is accomplished by external deflection coils, and additional fine focusing by the focusing and alignment coils. As beam electrons reach the photoconductive layer some are absorbed to

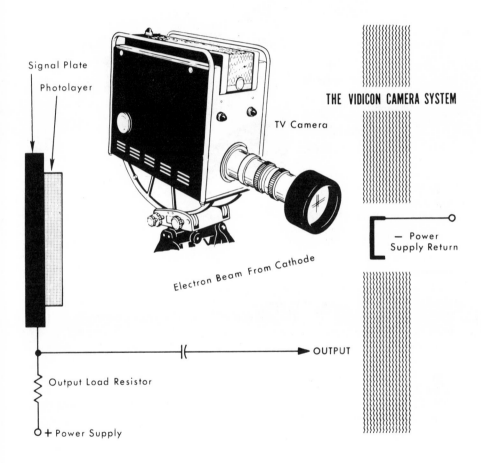

Signal Plate

Photolayer

THE VIDICON CAMERA SYSTEM

TV Camera

— Power
Supply Return

Electron Beam From Cathode

OUTPUT

Output Load Resistor

+ Power Supply

neutralize the positive charges that reside there due to the image, and excess electrons are discarded. The varying amounts of electron-absorption produce a current in the load resistor of corresponding intensity. The signal voltage drop across this resistor then serves as the camera output voltage, which is transferred through the coupling capacitor to the succeeding video amplifiers.

A racing car
caught by an
image orthicon.

The same car
picked up by
a vidicon is blurred
due to
slow speed response
or lag.

Other Vidicon Shortcomings

Despite the advantages of small size and simplified construction that characterize the vidicon tube, television producers are often reluctant to replace the image orthicon with it because of the impairment to picture quality that accompanies its use. We have already mentioned poor resolution and the need for greater scene illumination intensity, but there are several other drawbacks connected with the vidicon. Its speed of response, called "lag," under normal lighting conditions is so slow that pictures of moving objects are often blurred by smearing. Another undesirable characteristic of the vidicon is its relatively large dark currrent; this is the signal that appears even when there is no light incident on the tube. The principal effect of dark current is to produce uneven picture shading because the signal is not uniform over the entire area of picture scan and is also dependent on target voltage settings as well as on temperature.

Burn-in is frequently a severe problem in vidicon-equipped cameras, too. This effect is produced by prolonged exposure of the tube to bright stationary objects, sometimes leaving a permanent image in the photoconductive layer that necessitates replacement of the vidicon.

Yet, under certain field conditions the need for compactness and simplicity of accessory equipment has until very recently mandated the use of vidicons in spite of their shortcomings.

A Compact High-Performance Camera Tube

The recently developed *plumbicon* camera tube, only slightly larger than the standard 1-in. vidicon, has so successfully overcome the inherent weakness of the vidicon that it is now gradually replacing the latter in applications requiring portability and compactness without undue sacrifice of picture quality.

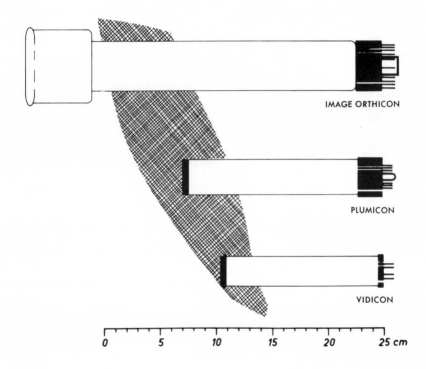

The plumbicon is closely related to the vidicon in basic principle. It utilizes the photoconductive properties of lead monoxide instead of the antimony trisulfide found in the vidicon. The lead monoxide is deposited in the form of a thin layer on a faceplate already carrying a transparent film of tin dioxide. The tin dioxide layer serves as the signal electrode. The lead dioxide is doped with impurities so that a kind of solid state sandwich having p-i-n characteristics is formed. That is, the layer on the electron gun side of the coating is made a P-type semiconductor while the layer on the signal electrode side is an N-type semiconductor. The middle layer is almost pure lead monoxide, thus forming an intrinsic semiconductor. The fact that the plumbicon can satisfy the stringent demands of broadcast television is largely the result of this special p-i-n diode structure of the photoconductive lead monoxide layer.

1-57

Advantages of the Plumbicon

The superior performance characteristics of the plumbicon contrasted with the vidicon are best appreciated by a listing as given below:

Vidicon	*Plumbicon*
Uneven, mottled picture shading due to large dark current.	Small dark current results in uniform background shading over entire scanned area.
Excessive picture smear due to lag under normal lighting conditions.	Even good vidicons retain up to 15% of a given initial signal after 3 scanned frames of the picture. Normal plumbicons coming off a typical production line average about 3% retention for the same time.
Relatively low sensitivity, requiring more intense lighting.	Sensitivity is considerably better. When compared under identical setting conditions, the plumbicon is capable of producing about 10 times the signal output.
Poor resolution compared to the orthicon.	Slightly better resolution at the corners.
Subject to damaging burn-in from bright stationary objects.	Burn-in is a minor problem. Any plumbicon that does not dissipate its burned-in image in 10 sec or less is rejected in manufacture.
Target blemishes appear and increase early in life of tube.	Target blemishes are due mostly to large dark currents. The plumbicon starts cleaner and stays cleaner than the vidicon over extended periods.

How a Magnetic Field Deflects an Electron Beam

The fact that magnetic fields can exert physical forces underlies the operation of many electrical devices.

Basic electricity teaches that a wire carrying current is encircled by magnetic lines of force which have a direction determined by the direction of the current. *(The direction of the magnetic field is based on the electron flow concept.)* The same principal applies to the electron beam. The electron beam is an electric current traveling through space (the vacuum in the tube) rather than through a conductor. It too is encircled by magnetic lines of force whose direction is determined by the direction in which the beam is propelled, i.e., the direction in which the electrons are moving.

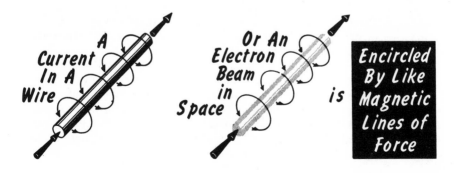

How a Magnetic Field Deflects an Electron Beam (cont'd)

There is a direct similarity in behavior between a current-carrying conductor located in a magnetic field and the electron beam which is projected through a magnetic field. In both instances the electron current is at right angles to the direction of its own magnetic field and we assume that it is at right angles to the externally produced field. An example of the current-carrying conductor in a magnetic field is found in the electric motor. An electron beam in a magnetic field is found in the cathode-ray tube typified by the iconoscope, the image orthicon, and the conventional television receiver picture tube (kinescope).

In the case of the electric motor, the external magnetic field is produced by current passing through coils wound on a core made of magnetic material; in the case of the cathode-ray tube the external magnetic field is produced by suitable currents passing through the deflection coils, only one of which is shown here.

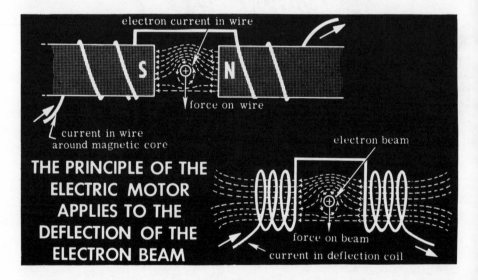

If you examine the directions of the lines of force of the two fields you will see that in each case they aid each other above the current and buck each other below the current. Where they aid each other the result is a strengthened field; where they buck each other below the field is weakened. Thus the stronger field is again above the current and the weaker field is below. When a current-carrying conductor or an electron beam is immersed in a nonuniform magnetic field, the field exerts a force which makes both the conductor and the beam from the area of the stronger field move toward the area of the weaker field. Hence both the conductor and the beam are pushed in the downward direction. The word "push," applied to the electron beam in a modern cathode-ray tube, means deflection of the beam.

Deflecting the Beam in the Horizontal Direction

Deflection of the electron beam in the iconoscope, in the image orthicon, or in any magnetically deflected cathode-ray tube is the result of a force acting on the beam in the horizontal direction and another force acting in the vertical direction at the same instant. To make this action more understandable, we shall deal with each direction separately, then combine them.

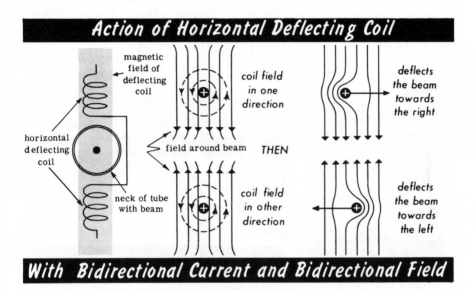

Action of Horizontal Deflecting Coil

magnetic field of deflecting coil

horizontal deflecting coil

field around beam

neck of tube with beam

coil field in one direction

THEN

coil field in other direction

deflects the beam towards the right

deflects the beam towards the left

With Bidirectional Current and Bidirectional Field

Deflection in the horizontal direction means motion of the beam *across* the mosaic, target, or picture tube screen from left to right, and then in the *reverse* direction. For this to happen the horizontal deflecting coil must be positioned *vertically*, at 90° relative to the direction in which the beam is to be moved. This relationship is dictated by the fact that the lines of force of the magnetic field of the coil are at right angles to the direction of the windings' turns. Also a factor is the lines of force of the field which encircle the beam being at right angles to the direction of advance of the electrons in the beam. When the two magnetic fields are oriented in this relationship, they combine to produce a *stronger* field on *one side of the beam than on the other*. Acting under the force exerted by the combined field, the beam moves in the horizontal direction, towards the weaker field.

Whether the beam moves towards the right or the left is determined by the instantaneous direction of the current flowing in the deflecting coil. To produce motion alternating between right and left, the deflecting-coil current is caused to alternately reverse its direction. This is a function of the device which generates the deflection voltage that is applied to the deflecting coil.

Deflecting the Beam in the Vertical Direction

The principle underlying the motion of the beam from side to side across the target also applies to the motion of the beam up and down along the surface of the target, except that the strengthened and weakened areas of the combined field now must be above or below the beam. For this to occur the orientation of the long axis of the vertical deflecting coil must be horizontal, at 90° relative to the direction in which the beam is to be moved. Then the lines of force from the coil current and the lines of force of the beam field will aid each other either above or below the beam, and the two fields will buck each other on the side opposite.

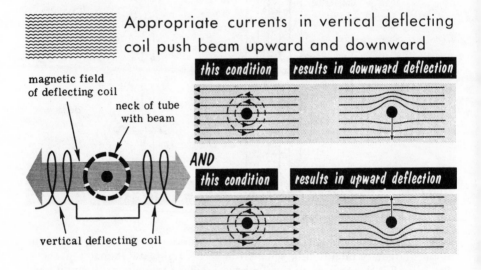

Appropriate currents in vertical deflecting coil push beam upward and downward

magnetic field of deflecting coil

neck of tube with beam

vertical deflecting coil

this condition

results in downward deflection

AND

this condition

results in upward deflection

Whether the beam moves upwards or downwards depends on the direction of the current flowing in the vertical deflecting coil. The direction of the beam also is a factor, but we assume its direction to be constant (from the electron gun towards the target to be scanned). We realize that the *return* beam in the image orthicon travels from the target toward the electron multiplier, which is opposite to the direction of the scanning beam. We do not, however, concern ourselves with the return when explaining the process of deflection as part of the scanning action. Let it suffice to say that given any direction of the deflecting coil field and a direction of the beam field, to change the direction of advance of the beam means changing the direction of its field and reversing the direction of deflection. What was previously to the left, would now be to the right; what was previously upward, would now be downward. To make the beam move up and down alternately, the vertical deflecting-coil current is reversed in direction alternately.

Simultaneous Horizontal and Vertical Deflection

We have said that the actual scanning of the mosaic or of any target by the electron beam is motion that represents the result of two forces acting simultaneously, one in the horizontal direction and one in the vertical direction. Just as an object cannot be in two places at one time, it can not move in two directions at one time. But it *can* move in a manner determined by two influences acting at the same time. What we mean when we say that the electron beam is deflected horizontally and vertically at the same time is that there is a motion which is the result of two components of force—a horizontal component and a vertical component—active simultaneously.

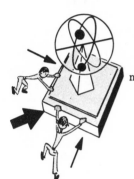

Two men pushing equally hard at right angles on a movable object will give the object forward motion in a RESULTANT direction

$F_1 = F_2$

You may recall learning in high school physics that two men pushing equally hard at right angles on a movable object will cause it to move in a direction that is the resultant of the directions of the two applied forces. If one man pushes with more force than the other man, the object will move forward in a new direction, *towards the direction in which the greater force is applied.*

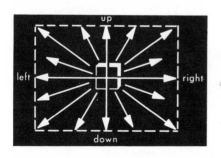

By controlling the direction and magnitude of the two component forces any given directional force can be achieved.
EVERY POINT ON THE RECTANGLE CAN BE COVERED.

The horizontal deflecting field and the vertical deflecting field are the equivalent of individual forces pushing the beam in two directions at right angles to each other.

Simultaneous Horizontal and Vertical Deflection (cont'd)

The behavior of two forces acting at right angles to each other, as explained on the preceding page, is directly comparable to the forces demonstrated by the horizontal and vertical deflecting fields acting together on the electron beam. The circular field around the beam, plus the horizontal and vertical deflecting coil fields, combine to aid or buck each other on two *adjacent sides* of the beam, rather than just on the left side, right side, above, or below when only one deflecting field is active.

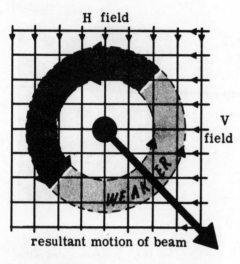

Combined H and V Fields Deflect the Beam at an Angle

Assume the H and V fields have the instantaneous directions shown. The three fields combine to produce a distorted field which is stronger *above and to the left* of the beam, and weaker to the *right and below* the beam. The beam is deflected toward the weakened field, which now is at an angle relative to straight right or straight down.

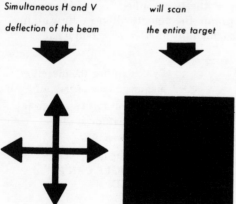

By changing the direction and instantaneous amplitude of the two deflecting currents we can control the direction and intensity of the resultant field, hence the direction in which the beam will move instant by instant, and also the distance that it will move. In this way we can deflect the beam so that it will touch every desired point on any target in a prescribed order.

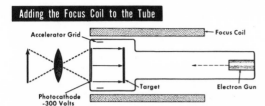

Adding the Focus Coil to the Tube

Accelerator Grid — Focus Coil — Electron Gun — Target — Photocathode -300 Volts

1. *The Image Orthicon* uses a low-speed scanning beam to increase its output by secondary emission. Electrons traveling from photocathode to target are kept in parallel paths by focusing coils which produce magnetic fields.

The Formation of the Charge Pattern on the Target

2. *The Target.* The side of the target facing the photocathode is struck by electrons. They are contained in isolated surface spots but are transmitted to the rear of the target so that the image appears there, ready to be scanned.

TRANSLUCENT PHOTOCATHODE WITH IMAGE OF WHITE ARROW — ELECTRON IMAGE — TARGET — CHARGE PATTERN OF IMAGE APPEARS ON BOTH SIDES OF TARGET

The + Signs Indicate Positive Charge

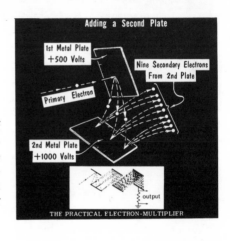

Adding a Second Plate

1st Metal Plate +500 Volts

Nine Secondary Electrons From 2nd Plate

Primary Electron

2nd Metal Plate +1000 Volts

output

THE PRACTICAL ELECTRON-MULTIPLIER

3. *The Electron Multiplier.* A series of dynodes (positive-voltage metallic surfaces) act as current amplifiers. The number of electrons they emit increases from dynode to dynode as each electron strikes and frees other electrons.

Two men pushing equally hard at right angles on a movable object will give the object forward motion in a RESULTANT direction

F_1 in this direction

resultant direction

F_2 in this direction

$F_1 = F_2$

4. *Deflection.* An electron beam is deflected from a stronger to a weaker magnetic field. The final beam direction results from the combined horizontal and vertical deflections.

Scanning

The deflection process positions the electron beam on the surface to be scanned. It may be the mosaic in the iconoscope, the target in the image orthicon, or the screen of the television receiver picture tube. In any case the beam describes a particular pattern of motion called the *scanning pattern* across the surface. In line with standards of television broadcasting employed throughout the world, the pattern used is *interlaced scanning.* This name denotes the way in which the televised image is reconstructed on the receiver picture tube screen, a process reversing the image's division into parts in the television camera and its subsequent transmission to the receiver as an electrical signal. Let us examine interlaced scanning.

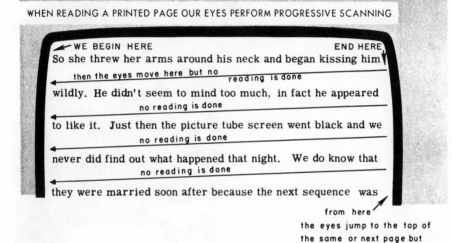

WHEN READING A PRINTED PAGE OUR EYES PERFORM PROGRESSIVE SCANNING

We can learn a great deal about the scanning process used in television by thinking of how we read a printed page. We read the printed page a line at a time, beginning at the left-hand edge of the page and moving towards the right. Having reached the end of the first line the eye sweeps to the beginning of the next line at the left side of the page, and the horizontal scan is repeated. Upon reaching the bottom of the page, the eye moves rapidly to the *left* edge of the *top* line of the *next* page.

What we have described is *progressive scanning*—reading one consecutive line after another. However, there are two intervals when no actual reading is done; while the eye is moving from the end of a line to the beginning of the next line, and while the eye is moving from the bottom of the page to the top of the next page. Bear these no-reading intervals in mind, because they occur in electronic scanning also.

How the Electron Beam Performs Interlaced Scanning

Progressive scanning is not used in television broadcasting. Interlaced scanning is used instead. Later on we show this phenomenon in detail, but for now all that is necessary for us to understand is the principle. The following example illustrates the idea of interlaced scanning.

The first page reads...

1	Once upon a time
2	There was a little girl
3	Named Red Riding Hood
4	Who lived with her grandmother
5	In a large forest

Let us assume that we have opened to the first page of a child's book in which there are five lines of words, also that this page is imaged on the target (or mosaic) of the camera tube, and finally that the target is divided into five horizontal rows of picture information. We will further imagine that these five rows are numbered 1 through 5. We know that there is no such arrangement in the actual camera tube, but what we have described will serve as a means of explaining the action.

Interlaced Scanning (cont'd) (The First Field)

The scanning beam in the camera tube starts moving across the target. Because of the design of the vertical and horizontal deflecting systems, and the voltages involved, the starting position of the beam is at the left-hand edge of the row of picture information. The first row corresponds to line 1 in our illustration. Because of the shape and the amplitude of the vertical deflecting voltage the horizontal excursions of the beam traverse rows 1, 3, and 5 of the image, skipping rows 2 and 4.

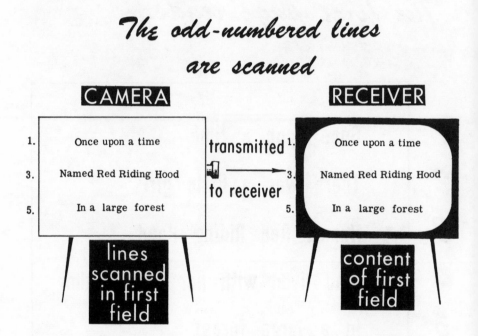

We have scanned the *odd-numbered* lines in the image and are now at the bottom edge of the target. The picture information as developed during the scanning of the odd-numbered rows is transmitted to the receivers bit by bit. Having completed the scanning of the odd-numbered lines in the image, the action is described as scanning a *field,* in this case the *odd field.* Also, having transmitted the picture information content of the odd-numbered lines to the monitor in the control room and to the receivers, we have transmitted the odd field to the receivers.

The picture information comprising the first or odd field appears on the screen of the receiver picture tube. How this occurs will be discussed later. In the meantime bear in mind that there is at present only a portion of the whole image on the camera tube mosaic.

Interlaced Scanning (cont'd) (The Second Field)

After completing the first field, the scanning beam is returned to the top of the target on which the complete image exists. But now, because of the starting position of the scanning beam, the even-numbered rows, that is rows 2 and 4, are scanned for picture information, while rows 1, 3, and 5 are skipped.

Scanning all the even-numbered lines (only two of them are used in this simple example) completes the scanning of the second, or even field. As in the case of the first field, the information developed bit by bit is transmitted as a train of signals. With the second field transmitted, the picture seen on the monitor screen and in the homes of the viewers is the complete image of the page as originally presented on the camera-tube target.

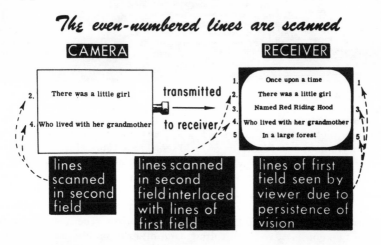

In the reconstruction of the picture (on the monitor or receiver picture tube screens) the even-numbered rows of the second field are interlaced between the odd-numbered rows which made up the first field. Space for the interlacing is provided for by the voltages present in the deflection system. The spaces skipped during the scanning of the first field are recreated on the receiving tube screens in between the odd-numbered lines of the picture that are reconstructed on the screen. The deflection voltages that move the beam during picture reconstruction have the same characteristics as the deflection voltages which account for the scanning in the camera tube, hence a row skipped during the scanning of the camera tube target results in a corresponding vacant row in the reconstructed image.

Observe that the scanning process does not disturb the location of the rows of picture information; it simply selects the odd-numbered rows for the first field and the even-numbered rows for the second field in the camera and places them in their correct relative positions on the receiver screen. It should be clear now, that the picture is sent in two parts.

Interlaced Scanning (cont'd) (The Frame)

A complete picture is seen *only* when two consecutive fields appear, interlaced, on the picture tube screen. When two consecutive fields have been transmitted, a *frame* has been transmitted. The presentation of the individual field content in terms of rows of picture information is illustrated below. To

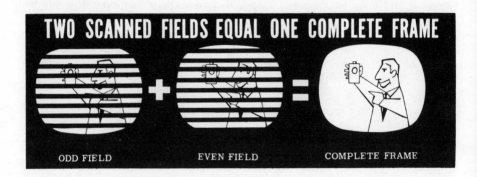

scan a complete field consumes 1/60 sec. This is true in the camera tube and at the receiver picture tube. Because a frame consists of two fields, the total time lapse for a frame is 2 × 1/60 sec., or 1/30 sec. These time intervals are very important for several reasons which are discussed later. In the meantime we should answer a question which no doubt has risen in your mind. If each field is transmitted separately how can the viewer see a complete picture corresponding to a frame? The answer is found in a characteristic of the human eye known as *persistence of vision*. The brain "memorizes" the first field, and because the two consecutive fields follow each other in such rapid sequence, the brain sees a complete frame or picture instead of two separate fields.

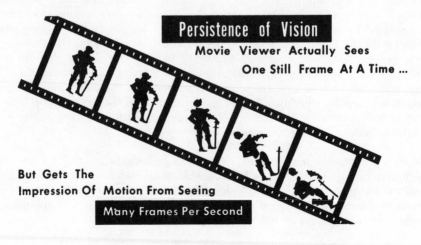

Technical Details of Interlaced Scanning

The concept of interlaced scanning illustrated by the child's book is easy to understand, but as we said before, only the principles described are technically accurate. The difference between the simplified explanation and the actual function is found in the number of rows of picture information that make up the picture, or more correctly the number of lines scanned in each field. Also, there are details relating to the motion of the scanning beam that have not yet been discussed.

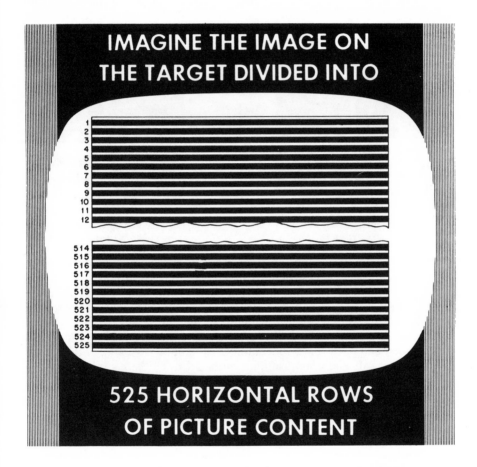

Whereas our original example of a printed page on the camera tube target was assumed to be scanned by only five horizontal excursions of the electron beam, the actual scanning of a camera tube target requires 525 horizontal excursions to complete a frame. The image on the target is divided into 525 rows of picture information.

Technical Details of Interlaced Scanning (cont'd)

We said that each image was transmitted in two fields, also that the first contained only the odd-numbered lines and the second only the even-numbered lines of picture information. Under these circumstances, the image made up of 525 lines of picture information is divided into two fields of 525/2 lines each. The beam scans 262.5 odd-numbered lines beginning with 1 and continuing through 3, 5, 7 . . . 521, 523, 525 completing the first field. The 262.5 even-numbered lines, theoretically beginning with line 2 and advancing through 4, 6, 8 . . . 520, 522, 524, are scanned by the beam completing the second field.

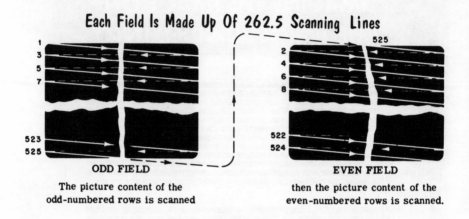

Each Field Is Made Up Of 262.5 Scanning Lines

ODD FIELD   EVEN FIELD

The picture content of the   then the picture content of the
odd-numbered rows is scanned   even-numbered rows is scanned.

The interlacing is made possible by specific positioning of the starting point of the scan of each field. The first field scan begins at the upper left-hand corner of the target and ends at the center of the lower edge. The scan of the second field starts at the center of the upper edge of the screen and ends at the lower right-hand corner of the target. The last scan of the first field is only a half-line, the first half of the line 525. On the other hand the first scan of the second field is also a half-line which can be considered to be the completion of line 525, except that it is at the top of the target rather than at the bottom. The first full line of the second field begins at line 2 of the complete image.

The vertical deflecting voltage accounts for the distance between the horizontal scanning lines described by the beam. It is the same kind of voltage for both fields, and by beginning the second field at the center of the top edge of the target rather that at the top left edge, the scanning lines of the second field fall in between the scanning lines of the first field, that is, they interlace. The importance of the interlace action occurs in the reconstruction at the receiving end, but the configuration of the scanning lines of the two fields at the camera is no less important, for what happens there influences what happens at the receiver, as will be seen later.

Technical Details of Interlaced Scanning (cont'd)

What does the interlaced pattern of scanning lines look like? We have shown two images produced by the interlaced scan of rows of picture information. However, neither one of these was really true to form. It is difficult to show the interlaced pattern of the scanning lines described by the beam while it is completing a frame because of the great number of lines involved. We can approach it by showing just some of the lines in their relative positions without showing any picture information. In other words only the pattern traced on the camera tube target (or on the picture tube screen) by the moving beam is illustrated. The solid lines show the left-to-right movement and the dashed

The Interlaced Scanning Pattern

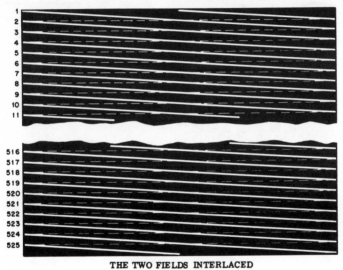

THE TWO FIELDS INTERLACED

Note: Instantaneous return from bottom to top of mosaic is assumed here.

lines show the right-to-left return of the beam. Note that the beam return from the right edge of odd-numbered lines is to the next odd-numbered line on the left. Also, the beam return from the right end of even-numbered lines is to the next even-numbered line on the left. The exceptions are the half-lines on the bottom and the top of the pattern. The dot-dash line showing the return of the beam from the bottom (at the end of the fields) to the top of the pattern makes horizontal excursions while advancing to the top. This is so because the horizontal deflecting voltage which accounts for the to and fro motion of the beam is always active. We will say more about this later.

Horizontal and Vertical Scanning Time and Frequencies

The horizontal motion of the scanning beam can be divided into two parts; the movement from the left edge to the right edge of the target (or screen), called the *horizontal forward trace* or simply *horizontal trace*, and the more rapid return from the right edge to the left edge. This is called the *horizontal retrace*. The horizontal trace time is 56 μsec (microseconds) and the horizontal retrace times is 7 μsec. The forward trace and the retrace action make up a horizontal cycle. The total time for the cycle is 63 μsec which corresponds to a frequency of 1/0.000063 or 15,750 Hz. The frequency of the horizontal deflecting voltage is therefore 15,750 Hz.

HORIZONTAL-VERTICAL SCANNING FREQUENCIES

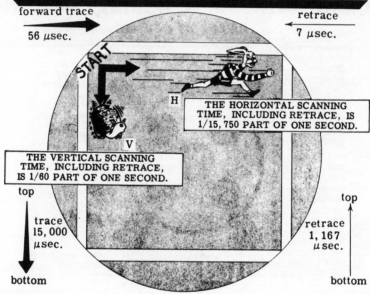

forward trace
56 μsec.

retrace
7 μsec.

H

THE HORIZONTAL SCANNING TIME, INCLUDING RETRACE, IS 1/15,750 PART OF ONE SECOND.

V

THE VERTICAL SCANNING TIME, INCLUDING RETRACE, IS 1/60 PART OF ONE SECOND.

top

trace 15,000 μsec.

top

retrace 1,167 μsec.

bottom

bottom

The vertical scanning action can be analyzed in the same manner. The downward action is the *vertical trace* whereas the return from the bottom edge to the top edge is the *vertical retrace*. The two motions complete a vertical scan cycle. The vertical trace time is 15,500 μsec and the vertical retrace time is 1167 μsec (allowing for tolerances). The sum of these two time intervals is 16,667 μsec, which corresponds to a frequency of 1/0.016667 or 60 Hz. Dividing 15,750/60 gives the number of horizontal scanning lines that are completed during one vertical cycle, namely 262.5 lines. Because the frequency references are simpler than the microsecond references, it is customary to refer to the horizontal and vertical scanning actions in terms of frequency rather than time elapsed per cycle.

Control of the Scanning Beam

The master timing device for controlling the entire television system, both transmitter and receiver, is the *pulse generator*. It produces five different kinds of control-voltage pulses, each of which performs a specific task relative to the control of the scanning beam in the camera tube. Two kinds of accurate frequency-control pulses are delivered by the pulse generator to the horizontal-vertical deflection-voltage generator. One determines the instant when the beam begins its horizontal forward trace and its retrace; another determines the instant when the beam begins its vertical deflection action downward and retrace after having completed a field. These control pulses are the horizontal and vertical sync pulses.

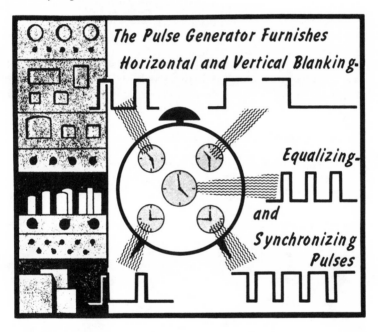

The Pulse Generator Furnishes Horizontal and Vertical Blanking, Equalizing, and Synchronizing Pulses

The two kinds of blanking pulses determine the beginning and the end of the periods when the camera tube ceases releasing picture information. The equalizing pulses are used to establish identical electrical conditions which enable the vertical sync pulse to initiate the vertical retrace at the correct moment in each field.

In addition to serving the camera tube, the pulses produced in the pulse generator are also delivered to the receiver for identically precise beam-control purposes in the receiver picture tube. To send each control pulse to the receiver, it is made part of the composite picture signal that eventually amplitude-modulates the carrier. The control pulses are added to the camera output signal in the control room and at the transmitter, as is explained later.

Horizontal and Vertical Blanking

When applied to an electron beam, "blanking" means beam cutoff which is similar to plate current cutoff in the conventional vacuum tube. In the camera tube or any other cathode-ray tube, beam cutoff prevents the electron beam from striking the mosaic, target, or screen, whichever it may be. The beam is stopped at the control grid of the electron gun by a sufficiently high negative bias applied to this electrode, or a sufficiently high positive bias applied to the cathode. Actually, the beam is extinguished.

Blanking creates action in the camera tube similar to the no-reading action the eyes experience while they move to begin a new line after reading the previous one, or while they are moving between a completed page and the top of a new page. Similarly, the electron beam is blanked when it has completed the scan of one line and is being retraced to begin the next horizontal scan. It is blanked again when the beam has reached the bottom edge of the mosaic or target and is returned to the top edge to begin the scanning of the next field, that is during vertical retrace. Failure to blank the beam would result in the development of confusing picture information during the horizontal and vertical retrace intervals.

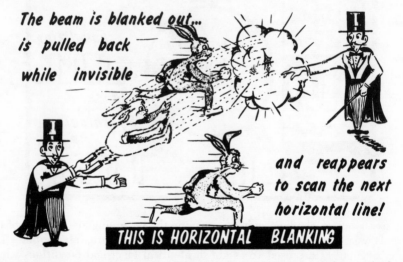

Horizontal blanking is accomplished by a blanking voltage pulse applied at a particular time in the sequence of horizontal scanning. It is a 10-μsec pulse that recurs 63.5 μsec apart, just shortly before the end of the scan of a horizontal line. It remains on for the 7-μsec retrace and for about another 2 μsec after the scanning of the next horizontal line has begun. Vertical blanking is the result of a blanking voltage that occurs once in every field. It lasts for about 1250 μsec and appears just about the time that the fourth from the bottom horizontal line scan begins, and lasts throughout the vertical retrace and over the period of the first 10 or 11 horizontal lines at the start of each field.

The Aspect Ratio

The proportions of the TV picture were set by the standards of the movie industry. The ratio of width to height of the traditional motion picture film is 4 to 3. If it is 4 ft wide, for example, it will be 3 ft high; if it is 16 ft wide, it will be 12 ft. high. The technical term for the relation of the width to the height is the *aspect ratio*.

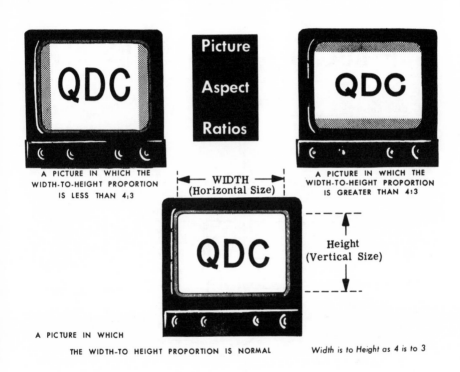

A PICTURE IN WHICH THE
WIDTH-TO-HEIGHT PROPORTION
IS LESS THAN 4:3

WIDTH
(Horizontal Size)

A PICTURE IN WHICH THE
WIDTH-TO-HEIGHT PROPORTION
IS GREATER THAN 4:3

Height
(Vertical Size)

A PICTURE IN WHICH
THE WIDTH-TO HEIGHT PROPORTION IS NORMAL

Width is to Height as 4 is to 3

The Aspect Ratio for the Receiver
Television receivers now in use have viewing screens of all sizes. Projection screens for use in theatres are just as large as movie screens. Yet, regardless of the size of the final picture, its dimensions must have the 4 to 3 ratio. The penalty for disregarding this standard is a distorted picture, because the aspect ratio is set in the television camera, and it is 4 to 3—width to height.

Horizontal and Vertical Size Controls
Hand controls for setting the width and height of the picture are provided in receivers. They are the size controls; vertical for the height of the picture, and horizontal for the width of the picture. As a rule they are located at the rear of the chassis.

Adding the Pulse and Deflection Voltage Generators to the Picture Signal Chain

Now add the pulse and deflection-voltage generators to the block diagram of the picture-signal chain. When doing this we must also add several related amplifiers. They are the horizontal and vertical deflection amplifiers which raise the amplitude of the respective deflection voltages produced in the deflection generator to the level required by the deflecting coils of the camera.

CONTROL OF THE SCANNING BEAM

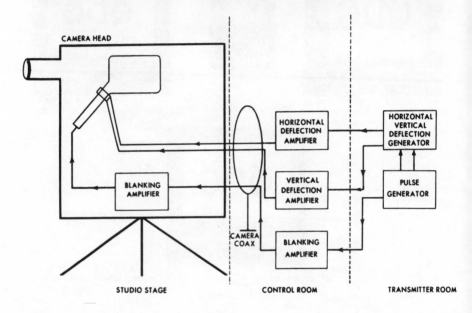

We also add the blanking amplifiers, one in the camera housing and one in the control room. The two amplifiers shape the blanking voltage pulses and raise their amplitudes to the required values. The pulse and deflection generators are in the transmitter room whereas the deflection amplifiers are in the control room. A coaxial (coax) cable containing the necessary conductors pipes the different control voltages to the camera tube. By properly locating the amplifiers and their controls in the control room it is easy for the control operators to comply with the desires of the director when he calls for special visual effects such as flop-over of an image or combining the output of two cameras into a single image. All of these effects relate to manipulation of the deflection voltages. The pulse generator is located where it can be under constant supervision by the engineers.

The Video Signal Amplifiers

In addition to the lines carrying the blanking and deflection voltages to each camera there are lines within the same coax cable which carry the picture information voltages from the cameras to the associated video amplifiers. We show only one camera and one video amplifier although usually there are three, four or even more. Each camera head contains its own video preamplifier to amplify the very weak video signal output of the camera close to the point of generation and to minimize the introduction of electrical noise into the camera signal.

Adding the Video Amplifiers to the Block Diagram

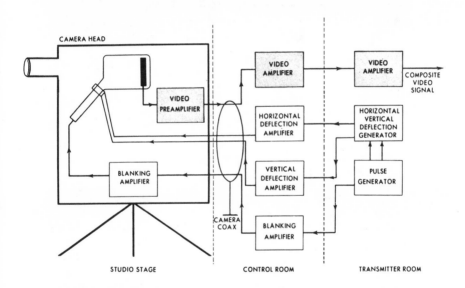

The video amplifiers in the control room are used to amplify the signals from the cameras and to feed signal voltages to the associated monitor picture tubes. The amplifiers also raise the picture signal amplitude to the desired level for delivery of the chosen picture to the transmitter. At the transmitter there is still another video amplifier which makes up for whatever loss in strength there occurred during transportation of the signal from the control room to the transmitter. It also acts to raise the picture-signal amplitude to the level required to amplitude-modulate the picture carrier. The output of the video amplifier in the transmitter room is the *composite video signal* containing picture and control elements, some of which are yet to be explained.

Adding the Monitors to the Video Chain

We have said that the output of each camera is displayed on an associated monitor in the control room for viewing by the director, allowing him to select the picture to be "put on the air." Let us add the monitor to the video chain. We show only one monitor with the one video amplifier in the control room. Many installations have four or five monitors and an equal number of video amplifiers.

ADDING THE MONITOR TO THE BLOCK DIAGRAM

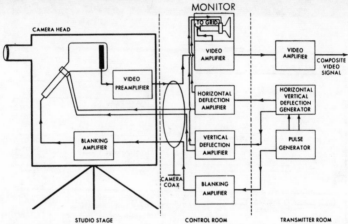

Each monitor is an assembly of a picture tube like one found in a home television receiver and components which enable it to function on the signal derived from its associated video amplifier. Each picture tube has its own electron gun for generating the electron beam, and horizontal as well as vertical deflecting coils to move the beam. The target for the beam is the inside surface of the glass envelope at the large end of the tube. This is the screen, with a surface coat of special material that glows or fluoresces when the beam strikes it. Whereas the camera tube beam acts to convert light energy into electrical energy (picture information voltages) the electron beam in the monitor tube receives its picture information voltages from the video amplifier and converts it into light energy. It "paints" the camera image on the screen, using tiny areas of light and dark spots caused by the beam striking the screen.

To make certain that the monitor beam moves in exact synchronism with the beam in the camera tube, the deflection voltages applied to the camera tube are also applied to the monitor-tube deflecting yoke windings.

The Video Signal from the Iconoscope Camera

Having assembled the vital elements of the picture channel in the transmitter, let us now examine the picture signal output from the camera. Since both the iconoscope and the image orthicon are still being used as television camera tubes we shall examine the action of each. (The block diagrams of the picture channel of the transmitter will continue showing the iconoscope.)

THE VIDEO SIGNAL PRODUCED BY SCANNING
ONE LINE OF AN IMAGE IN THE ICONOSCOPE

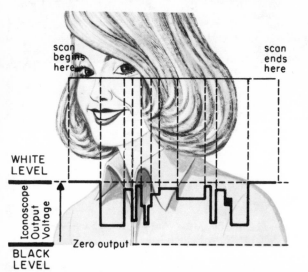

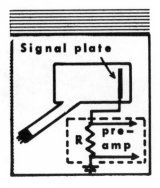

The iconoscope output current is fed into the camera tube preamplifier.

The Iconoscope

Our picture subject is a pretty girl. We know that the amplitude of the output signal is a function of the amount of illumination which has fallen on the photodroplets that make up the mosaic. What does the camera tube output signal look like? To find the answer let us scan a single horizontal line through the face of the subject. Maximum voltage output corresponds to white in the image; minimum voltage corresponds to black in the image; zero reflectance, or jet black, which produces zero output voltage, apparently is not present in the line scanned. Values of gray, between white and black, result in output voltage amplitudes between maximum and minimum. The abrupt changes in output voltage shown by the steep sides of the voltage wave are based on abrupt changes in the light and dark areas along the line being scanned. More gradual changes in shading would result in less abrupt fluctuations in output voltage. We should mention here that the usual method of illustrating line by line picture voltages derived from a camera tube is more often symbolic than realistic. We are using a representational drawing here only to facilitate our understanding.

The Video Signal from the Iconoscope Camera (cont'd)

The voltage equivalent of any horizontal line that is scanned begins where the beam starts scanning the image horizontally and ends where the scanning finishes and the horizontal retrace blanking begins. If we assume that the scanning action continues line after line, as happens in television transmission, we find we have delivered to the video preamplifier (in the camera) a *train* of picture voltages with intervals of blanking between.

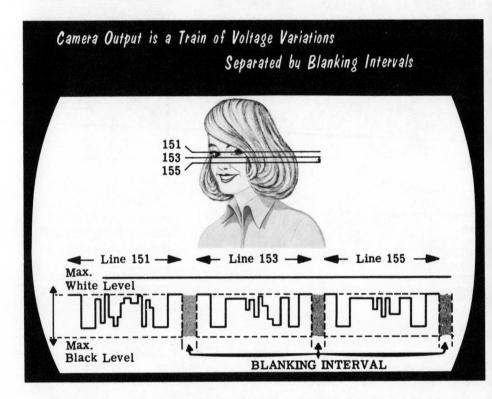

Our illustration shows the video output for only three consecutive odd-numbered lines that were scanned in a field, say lines 151, 153, and 155, but several important conditions relating to the entire image were brought to light by these three lines. Although the lines scanned are positioned one below the other, the voltage representation of these lines is a *train* of fluctuations, one following the other, along a time base. A blanking voltage is applied to the camera tube to extinguish the beam during horizontal retrace, although the signal output from the camera is zero during the blanking interval. In other words the blanking signal fed to the electron gun of the camera tube does not appear in the camera tube video output. The picture voltage representation of every line scanned is separated from the next by the blanking interval.

The Video Signal from the Iconoscope Camera (cont'd)

We have shown that the picture signal output from the iconoscope increases in amplitude as the image changes from black to white, and that white in the image produces the stronger signal. But this is not the way the signal is processed. The American standard of television transmission utilizes negative video polarity. This means that the conditions of operation call for white in the image to produce the theoretically zero signal and black in the image to produce the theoretically maximum picture signal.

HOW ICONOSCOPE SIGNAL OUTPUT IS GIVEN NEGATIVE VIDEO POLARITY

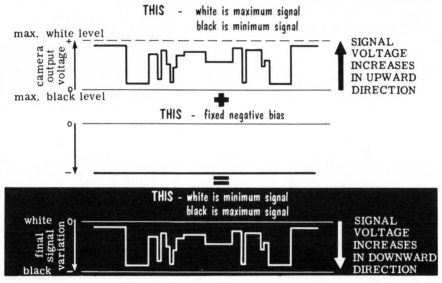

This requirement can be satisfied using the iconoscope camera by simply applying the camera output signal to a negative clamping circuit. Then the entire camera signal appears below the zero-voltage base line and is entirely in the zone of negative polarity voltages. When the fixed negative bias is added to the positive-polarity camera signal the result is a signal-voltage variation wherein white in the image becomes the least negative voltage and black in the image becomes the most negative voltage. This is readily understandable if we use a few simple figures as examples. Assume that the camera output for white in the image is +1.9 V and that black in the image produces an output voltage of +0.1 V.

If we add these voltages to a fixed negative bias of 2 V, then the final fluctuation in signal voltage is between −0.1 V (white in the image) and −1.9 V (black in the image). Thus black is the higher signal voltage.

The Video Signal from the Image Orthicon

The picture signal output from the iconoscope has been described as maximum for white areas in the image and minimum for black areas. How does this compare with the signal output from the image orthicon? Let us assume that the image and the line being scanned are the same as before. The charge pattern developed on the target has the *maximum* charge for the white areas and *minimum* charge (theoretically zero) for the black areas in the image. Relating these conditions to the return beam, we find it contains fewer electrons (less current) when returning from a white-area charge on the target and the normal amount of electrons (maximum current) when returning from a black-area charge on the target. Converting these currents into voltage under normal circumstances, white in the image produces minimum output voltage and black in the image produces maximum output voltage. The result is a negative-polarity video signal from the image orthicon.

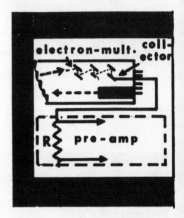

THE VIDEO SIGNAL PRODUCED BY SCANNING ONE LINE IN THE IMAGE ORTHICON

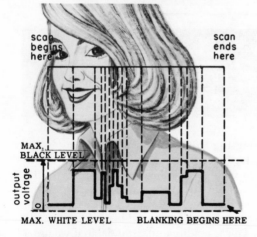

The return beam current after amplification in the electron multiplier flows through the camera tube load resistor.

The difference in behavior of the iconoscope and the image orthicon is for us only of academic interest. We are concerned mainly with the control of the scanning beam in each camera tube and with the content of the television picture signal that is radiated to the receiver. Like the signal output of the iconoscope, the image orthicon signal is picture information only; blanking as well as the other control voltage pulses must be added to the picture information in correct time sequence.

Adding the Blanking Pulses to the Picture Signal

To display the picture on the monitor screen and eventually on the receiver tube screen without showing the horizontal and vertical retrace motions of the beam, horizontal and vertical blanking voltages must be added to the picture signal. These pulses must appear in the same time sequence as the horizontal and vertical retraces during scanning. To assure perfect timing, blanking voltages are supplied from the pulse generator through the blanking amplifier in the control room, the same source which supplied blanking voltages to the camera tube. A cable connection between the blanking amplifier and the video amplifier in the control room feeds the blanking pulses from one to the other.

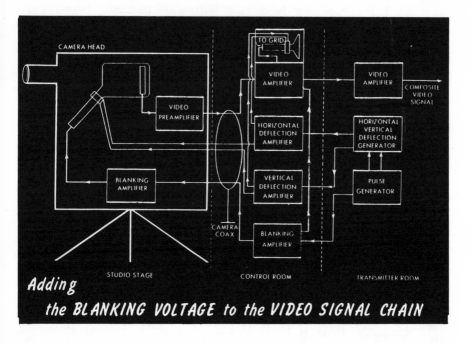

Adding
the *BLANKING VOLTAGE* to the *VIDEO SIGNAL CHAIN*

The picture signal and the blanking voltages are combined into a train of signals in the video amplifier. Then the combination signal containing picture and blanking information is fed to the monitor tube. The same combination signal is also available for delivery to the video amplifier at the transmitter. Whether or not it is sent there for additional processing (as will be described later) is determined by the director who selects the picture that is put on the air. As we stated earlier, each studio camera in use on the stage delivers a picture signal to its video amplifier and related monitor in the control room, thus allowing the director to select the angle shot he chooses.

Adding the Blanking Pulses to the Picture Signal (cont'd)

The blanking pulses are of negative polarity. Their amplitude varies between zero and a suitable peak value which exceeds the peak amplitude of the signal corresponding to black in the image. Also, they are of two different durations: the horizontal blanking pulses have a duration period that is only a small fraction of that of the vertical blanking pulses. The pulses are generated as a continuous train and follow each other in the same sequence that patterns the beam retrace intervals during scanning. Since there are many more horizontal than vertical retrace intervals in a field, there are many more horizontal than vertical blanking pulses in the train. In this regard we should clarify a point. Horizontal blanking in a field begins after the first 7 to perhaps 12 horizontal lines have been scanned, and ends when about 2 to 4 horizontal lines remain to be scanned. The reason for the delay in starting the horizontal blanking and for ending it before the field is completed is the vertical blanking pulse. Because of its time duration it starts before the beam is at the bottom of the camera tube mosaic (or target). That is when about 2 to 4 horizontal lines remain to be scanned in each field. It is active during the entire vertical retrace and remains active during the scaning of the first 7 to 12 lines of the next field. Starting early and ending late is insurance that the much slower vertical retrace will not be visible on the picture tube screen. Picture information is not delivered by the camera tube while vertical blanking is active. Adding the blanking voltage pulses to the picture signal fills the gap, labeled earlier the no-picture-signal-interval, between the end of one horizontal line of picture information and the beginning of the next line. We illustrated the results of

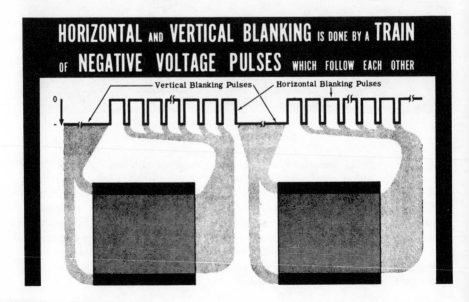

HORIZONTAL AND VERTICAL BLANKING IS DONE BY A TRAIN OF NEGATIVE VOLTAGE PULSES WHICH FOLLOW EACH OTHER

Vertical Blanking Pulses — Horizontal Blanking Pulses

Adding the Blanking Pulses to the Picture Signal (cont'd)

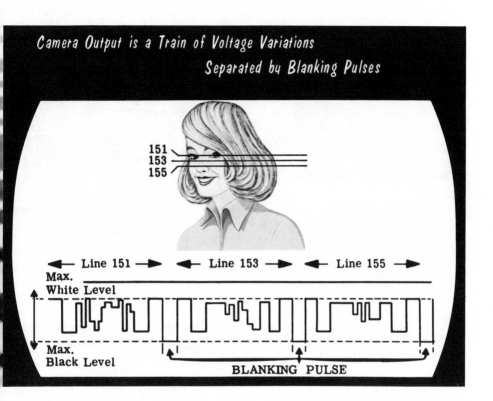

Camera Output is a Train of Voltage Variations Separated by Blanking Pulses

151
153
155

◄— Line 151 —► ◄— Line 153 —► ◄— Line 155 —►

Max. White Level

Max. Black Level

BLANKING PULSE

only three lines of picture information. Here we show the same three lines with the blanking pulses added. Continuous scanning of field after field results in a signal consisting of an unending series of picture information voltages followed by blanking voltage.

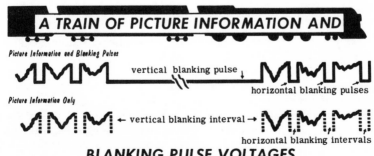

A TRAIN OF PICTURE INFORMATION AND

Picture Information and Blanking Pulses

vertical blanking pulse ↓

horizontal blanking pulses

Picture Information Only

← vertical blanking interval →

horizontal blanking intervals

BLANKING PULSE VOLTAGES

Details of Horizontal Blanking Pulses

Before concluding the discussion of blanking pulses, it might be well to detail the constants and action of the horizontal blanking pulse. The conditions of its use are of special interest, and are not evident simply by observation of the scanning lines that make up the fields.

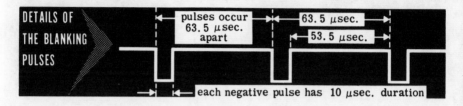

In line with the constants shown above the distribution of the horizontal blanking time is the following:

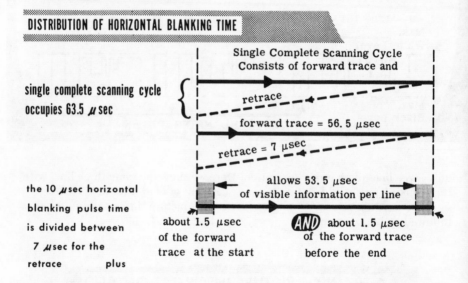

The blanking of the horizontal scan for the short intervals after it begins and before it ends is not the same in the camera tube as in the monitor and receiver picture tubes. The horizontal blanking pulse delivered to the camera tube is slightly shorter in duration than that which is combined with the signal. This is a safety measure to ensure that the beam is in the proper position before the delivery of picture information begins. The slight loss in picture content because of the extended horizontal blanking time is inconsequential, as is the picture content lost by the complete blanking of a number of horizontal scanning lines.

ODD FIELD   EVEN FIELD   COMPLETE FRAME

1. *Interlaced Scanning.* All odd-numbered lines are scanned as a group (a field), then all even-numbered lines. The first and last lines of each field are only half-rows; interlacing begins here. The complete inerlaced scan of two consecutive fields (the first held in the viewers eye by persistence of vision) produces an intelligible picture.

SCANNING ONE LINE IN THE IMAGE ORTHICON

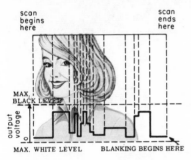

scan begins here   scan ends here

MAX.
BLACK LEVEL

output voltage

MAX. WHITE LEVEL   BLANKING BEGINS HERE

2. *Output Voltages in the Iconoscope* are determined by the degree of illumination on the mosaic. The voltages from a series of scanned lines are broken by the blanking pulses.

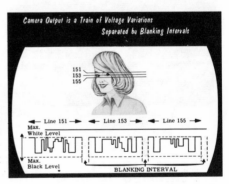

Camera Output is a Train of Voltage Variations
Separated by Blanking Intervals

151
153
155

◄— Line 151 —►   ◄— Line 153 —►   ◄— Line 155 —►

Max.
White Level

Max.
Black Level
BLANKING INTERVAL

3. *Comparative Video Signals.* The strength of iconoscope voltages increases as the image becomes whiter. In the image orthicon, an applied negative bias causes the voltages to appear in such a manner that black produces the maximum voltage.

4. *Blanking Voltages* are supplied from the pulse generator and, together with the picture signal, pass to the video amplifier. There are separate horizontal and vertical blanking pulses. While they are active, no picture information is transmitted.

Picture Information and Blanking Pulses

vertical blanking pulse

horizontal blanking pulses

Picture Information Only

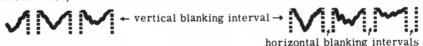

◄— vertical blanking interval —►

horizontal blanking intervals

Adding the Sync Pulses to the Video Signal

Reconstruction of a television scene on the screen of a monitor or receiver picture tube requires that the electron beam in this tube describes exctly the same horizontal and vertical motions as does the electron beam in the camera tube. It is not enough for the two beams to move under the influence of deflecting currents of identical frequency; the current variations must be perfectly in step with each other.

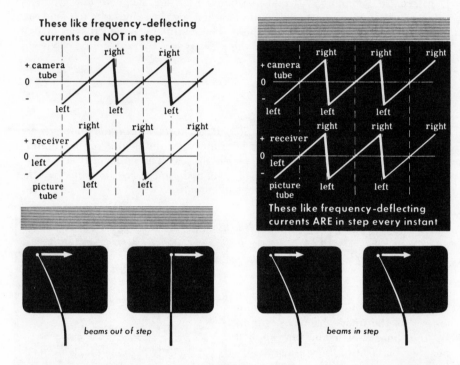

beams out of step   beams in step

BEAM MOTIONS IN CAMERA AND PICTURE TUBES

The two beams must change direction at the same instant, and must reach their minimum and maximum values in step with each other. If the two beams do not move in synchronism an unintelligible picture will be reconstructed. To ensure perfect reconstruction of the televised scene, the vacuum tube generators that supply the horizontal and vertical deflection voltages to the deflecting coils in the camera, and those in the receiver which produce the deflection voltages for the deflecting coils of the picture tube, are synchronized by voltage pulses derived from the same source—the pulse generator at the transmitter.

Adding the Sync Pulses to the Video Signal (cont'd)

The horizontal and vertical deflection voltages which cause the corresponding deflecting currents in the camera tube deflecting coils originate in the deflection generators. The deflecting voltages generated in the receiver and fed to the deflecting coils which serve the receiver picture tube must be not only of the same frequency as those active in the camera tube, but must also vary in step moment by moment. To ensure this, deflecting voltage sources in

Adding Sync Pulses
To The Video Signal

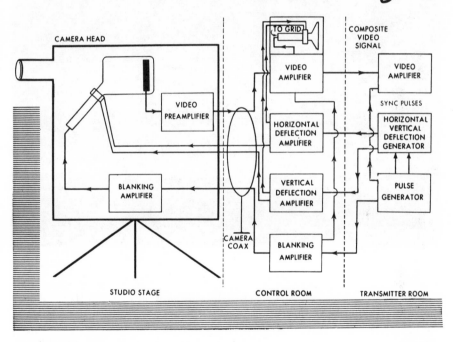

the receiver are placed under the control of the pulse generator in the transmitter. The timing or synchronizing pulses which control the deflection voltage generators are made a part of the picture signal that is sent to the receiver. A cable from the pulse generator feeds horizontal and vertical sync pulses to the video amplifier at the transmitter. It is in this amplifier that the sync pulses are added to the picture signal which now also contains the blanking voltage pulses.

Adding the Horizontal Sync Pulse

Sync pulses are of two kinds, horizontal and vertical. At this time we shall discuss only the synchronizing pulse that is related to the horizontal movement of the beam.

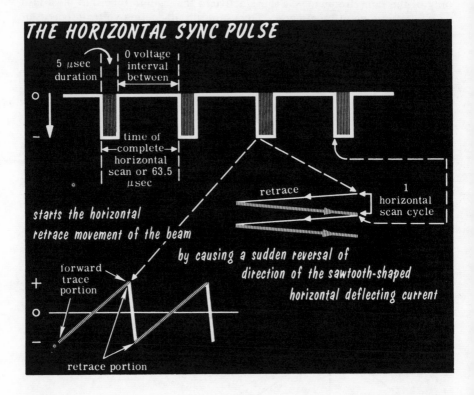

The horizontal sync pulse is very narrow; it lasts for only 5 μsec. Its polarity is the same as that of the horizontal blanking pulse, for as will be seen shortly, the sync pulse rides on the blanking pulse enabling its separation from the picture signal in the receiver. The pulse is timed to appear coincidentally with the arrival of the beam at the end of its horizontal excursion towards the right. It triggers the horizontal deflection generating system to initiate the retrace portion of the horizontal scanning cycle. In other words the horizontal sync pulse appears at the end of each horizontal scan. In this way the electron beam in the receiver picture tube is positioned at the proper point on the screen to accept the picture information being developed in the camera tube. We should emphasize that the destination of the horizontal sync pulse after delivery to the receiver is the horizontal deflection-current generator system.

Adding the Horizontal Sync Pulse (cont'd)

Each horizontal sync pulse is timed to appear just slightly after the start of the 10-μsec horizontal blanking pulse which appears at the end of each horizontal scan. Having a duration of only 5 μsec, the sync pulse ends before the blanking pulse. Therefore, the sync pulse occurs within the duration period of the blanking pulse. By making the peak amplitude of the sync pulse greater than

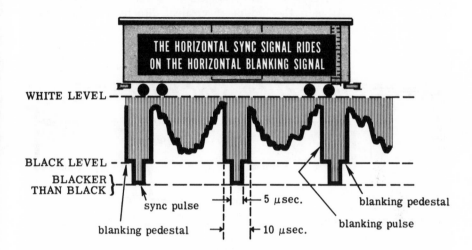

that of the blanking pulse, the sync pulse extends beyond the blanking pulse pedestal when the two are seen together. This makes it possible to "clip" the sync pulse off the blanking pulse pedestal in the receiver and to use it as a timing signal. Also, because the sync pulse extends beyond the black level set by the blanking signal pedestal height, the sync pulse signal is said to be located in the "blacker than black" region of the picture signal.

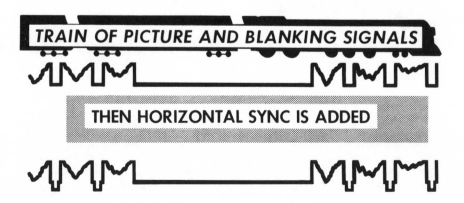

Adding the Vertical Sync Pulse

It is not enough that the receiver picture tube beam move from side to side in synchronism with the camera tube beam. The two beams must also move in step in the vertical direction. To make this happen is the function of the vertical sync pulse, which, like its horizontal counterpart, is used in the transmitter, and is received as part of the picture signal. In both the transmitter and the receiver the vertical sync pulse is applied to the vertical deflection generator where it triggers the sudden reversal of direction of the vertical deflection current. In so doing it starts the scanning beam moving upward at a

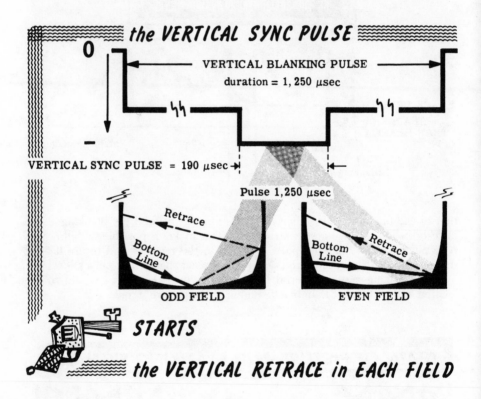

particular moment in relation to the overall scanning. The vertical retrace action of the scanning beam occurs once in each field; when the beam, having scanned the bottom horizontal line, is ready to move to the top of the scanned surface and begin scanning the next field. The vertical sync pulse appears every 1/60 sec or 60 times per second. The pulse interval is 190 μsec. This makes it a long-duration pulse in comparison with the 5-μsec horizontal pulse.

Adding the Vertical Sync Pulse (cont'd)

The beam moves upward, but not straight up; it zig-zags from side to side because of the action of the horizontal deflecting field which is ever present in both the camera tube and in the receiver picture tube. This movement must occur if the beam is to start scanning horizontally the moment the vertical retrace is completed. In this connection, each beam must be at its correct location—the camera tube beam ready to start the development of the picture information for transmission and the receiver tube beam ready to reconstruct the picture. For both the camera tube and the receiver picture tube, the moment *when* retrace begins in each field is extremely important. Each must begin at one precise instant if correct interlace of the two fields is to be accomplished. For this reason, and also to make certain that the horizontal syncing is continuous, the vertical sync pulse is notched or serrated; another train of pulses called the equalizing pulses is added ahead of and after the vertical sync pulse.

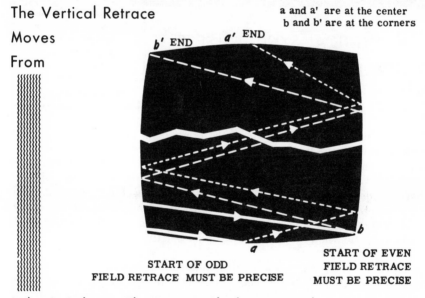

The Vertical Retrace

Moves

From

a and a' are at the center
b and b' are at the corners

b' END   *a'* END

a

START OF ODD
FIELD RETRACE MUST BE PRECISE

START OF EVEN
FIELD RETRACE
MUST BE PRECISE

Side to Side As The Beam Climbs Upward

You will recall the statement that there are 262.5 horizontal lines in each field. The simplified presentation of the odd and even fields showed the odd lines numbered from 1 through 525 and the even lines numbered from 2 through 524. Obviously the vertical retrace period of each field must consume some of these lines. This happens, but it is not necessary to again present the organization of the lines in each field. We simply accept the fact that about 20 lines per field are inactive because of vertical blanking.

The Serrated Vertical Sync Pulse

When thinking about synchronizing the vertical deflecting system of the receiver and the camera tube, we should not lose sight of a similar need for synchronization in the horizontal deflecting system. The simplified version of the vertical retrace shows it as an instantaneous movement from the bottom edge to the top edge of the mosaic, screen, or target. Actually it does not retrace this way.

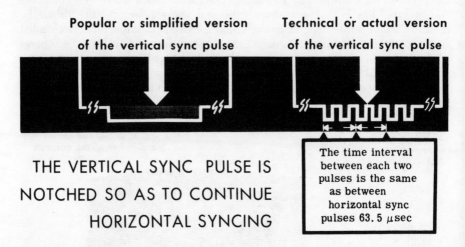

Popular or simplified version
of the vertical sync pulse

Technical or actual version
of the vertical sync pulse

THE VERTICAL SYNC PULSE IS
NOTCHED SO AS TO CONTINUE
HORIZONTAL SYNCING

The time interval
between each two
pulses is the same
as between
horizontal sync
pulses 63.5 µsec

To maintain horizontal sync during the presence of the vertical sync pulse in each field, the pulse is notched or serrated. This reforms the single long pulse that we illustrated into six like pulses of comparatively long duration with short intervals of no sync voltage between. The serrations are positioned so that the time interval between the leading edges of each two adjacent notches is the same as between two adjacent horizontal sync pulses. In this way the horizontal deflection generator is subjected to three horizontal sync impulses during the time when the vertical sync pulse is active in each field.

The vertical sync pulse on the vertical deflection generator is not impaired by the short intervals when the sync voltage falls to zero (the notches). Vertical syncing behaves as a constant amplitude pulse. As will be shown later in the course, the receiver system distinguishes between the horizontal and the vertical sync pulses due to the difference in duration and voltage intervals between these pulses. (You can see this difference for yourself if you compare the above notched vertical sync pulse with previous illustrations which show the horizontal sync pulses.)

Serrating the vertical sync pulse does not in any way change its relationship to the blanking pulse on which it rides as illustrated in the preceding drawing. Everything said before still applies.

Completing the Vertical Sync Action (Equalizing Pulses)

The description of the vertical sync action is completed when we add the *equalizing pulses*. They appear as two trains of six pulses each in each field; one train timed to appear just before the vertical sync-pulse interval and one train immediately after. They have the same polarity as the vertical sync pulse, but are unlike the latter in that they are short duration pulses. They, too, are formed so that the interval between two adjacent pulses corresponds to a horizontal scanning cycle, thus providing the equivalent of three horizontal sync pulses immediately before the vertical retrace appears, and three horizontal sync pulses immediately after the vertical sync pulse. In this way they keep the horizontal deflection generator on frequency during the period when the vertical deflection generator is being made ready for the reversal of the direction of the deflection current. This is the initiation of the vertical retrace.

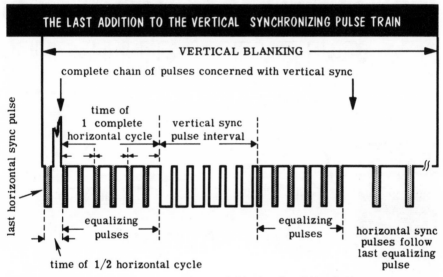

Note. these are the equalizing pulses for one field only - the odd line field

The "get ready" interval is required because each field contains a half-line. The instant of triggering of the vertical retrace action of the odd line and the even line fields differs by a half horizontal line. The last horizontal line in the odd field ends at the center of the bottom edge of the surface being scanned, completing the last horizontal line of the even field which ends at the right corner of the bottom edge. The equalizing pulses in each field take care of this time difference and trigger the vertical retrace of the successive fields at the proper moments so that correct interlace of the horizontal scanning lines takes place.

The Composite Video Signal

The addition of the horizontal and vertical blanking pulses, the vertical sync pulses, and the two trains of equalizing pulses to the picture signal produces what is called the *composite video signal*. This is the signal output from the video amplifier at the transmitter, and is used to amplitude-modulate the picture carrier. Since the process of composite signal transmission is continuous as long as the transmitter is on the air, any single train of signals may represent a short period of transmission without regard to the field, or portion of the field, that is being transmitted, unless the contrary is specifically indicated. We should also mention that it is customary to symbolize the waveform of the picture content between horizontal blanking pulses by the use of jagged lines.

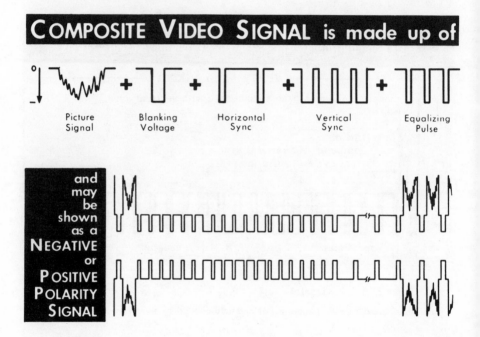

Another point which should be mentioned is the polarity of the composite video signal. Our discussion so far has been concerned with negative polarity signals although the illustration of the composite signal is more often one of positive polarity. The sync pulse peaks are the highest signal levels and the white content of the picture is the lowest signal level. However, the lowest peak signal level is not equal to the zero level of the signal used for modulation. The FCC and industry standards set the whitest signal at 5 to 10% higher than the camera's white output.

The Generation of the Picture Carrier

Before the composite video signal can be radiated from the transmitting antenna, it must be applied to a suitable RF carrier to produce an amplitude-modulated picture signal. This AM picture signal containing the video information, blanking pulses, sync pulses, and equalizing pulses is radiated from the transmitting antenna to be received by any TV receiving antenna in the viewing area.

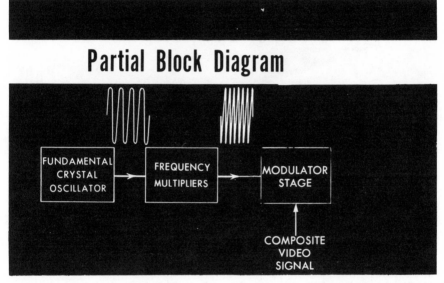

Partial Block Diagram

Like the sound carrier described earlier, the picture carrier is either in the very high frequency (VHF) or in the ultra high frequency (UHF) region of the electromagnetic spectrum, as determined by the operating picture carrier frequency allocated to the television station by the Federal Communications Commission. Also, as in the case of the sound carrier, the allocated picture carrier-frequency voltage cannot be generated directly.

The required degree of accuracy would be difficult to reach. Therefore, an accurately controlled signal is generated that has a much lower frequency than the final picture carrier frequency, but is a submultiple of the final carrier frequency. Then by a series of frequency-multipliers the signal frequency is raised to the required figure, at which time the amplitude-modulation is applied. We have described the process of frequency-multiplication in connection with the sound carrier, and shall not repeat it here. However, we must emphasize that the picture carrier is amplitude-modulated; therefore there is no frequency-modulation present at either the stage where the basic frequency signal is produced or anywhere along the chain of frequency-multipliers. The modulated carrier varies in amplitude in accordance with the instantaneous changes in the amplitude of the modulating signal, which is the composite video signal.

The Generation of the Picture Carrier (cont'd)

With the multiplier stages performing the task of raising the frequency of the picture carrier to the final figure, a number of stages of amplification are needed to raise the amplitude of the carrier to the proper level for modulation. This is performed in a series of radio frequency amplifiers.

Adding RF Amplifiers to the Picture Carrier Block Diagram

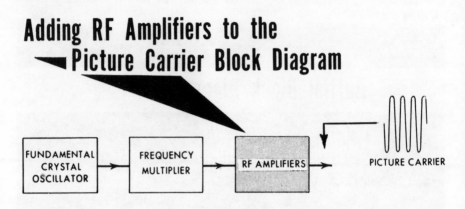

The composite video signal must be combined with the RF picture carrier. This combining occurs in the modulator stage. It is an RF amplifier to which are fed the carrier signal from the RF amplifiers and the composite video signal from the video amplifier at the transmitter. The output of the modulator is an RF signal that varies its amplitude in accordance with the amplitude variations of the composite video signal. It has an upper and a lower sideband just as any amplitude-modulated carrier has. Now the modulated signal has a frequency band width which is equal to twice the highest frequency present in the composite video signal.

ADDING THE MODULATOR STAGE TO THE BLOCK DIAGRAM

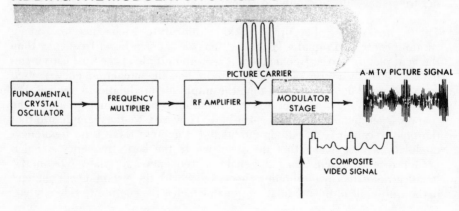

The Power Amplifier

The amplitude-modulated RF carrier signal is fed into a number of power amplifiers, which in turn feed the signal to the transmitting antenna for radiation to the receivers. These amplifiers perform a dual function in many transmitters. In all cases they raise the level of the modulated carrier to the amount required for transfer to the antenna and the attainment of the desired radiated power. The effective radiated power rating of a television transmitter is a function of the energy level obtained from the power amplifiers and also of the power gain achieved in the antenna. The meaning of power gain is explained elsewhere in this course.

The Block Diagram
of the Picture Part of the T V Transmitter

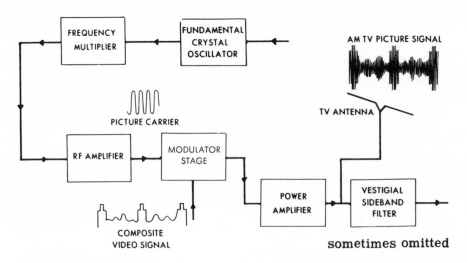

The second function performed by the power amplifiers in some transmitters is to limit the extent to which the lower sideband of the modulated picture carrier is transmitted. The amplifiers are operated somewhat off-tune. This allows the carrier and the entire upper sideband, but only about 1.25 MHz below the carrier frequency of the lower sideband, to be fed to the antenna. In some transmitters the power amplifier output contains the full upper and lower sidebands, but the lower sideband frequencies beyond 1.25 MHz are removed in special filter circuits that are interposed between the output of the power amplifiers and the antenna. They are known as *vestigial* sideband filters. This picture-carrier treatment conforms with vestigial sideband or partially attenuated sideband form of transmission.

The TV Picture Signal Sidebands

The usual amplitude-modulated signal is described as having a carrier, an upper and a lower sideband. The sidebands contain all the picture signal intelligence. The two sidebands are alike in bandwidth, the maximum bandwidth determined by the highest frequency component of the modulating signal. For instance, if the highest frequency component of the modulating signal is 7.5 kHz then the overall bandwidth of the amplitude-modulated signal is 15 kHz or twice the highest frequency component of the modulating signal. If the carrier is 600 kHz, the upper sideband signal frequency limit is f_c + f_m or 600 + 7.5 or 607.5 kHz. The lower sideband signal frequency limit is f_c − f_m or 600 − 7.5 or 592.5 kHz.

BANDWIDTH of the TELEVISION PICTURE SIGNAL BEFORE CORRECTION

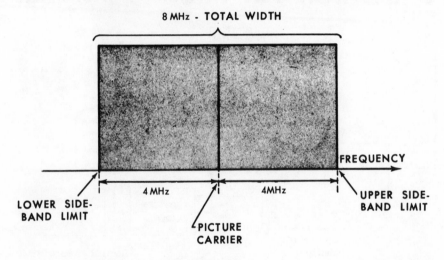

Except for the numerics involved, the above description fits the amplitude-modulated television picture signal at the output of the modulator in television station transmitters, and at the output of the power amplifier in some instances. American standards of television transmission set the highest frequency component of the composite video signal at 4 MHz. Under the circumstances the amplitude-modulated picture carrier has two sidebands, each 4 MHz wide, for a total signal bandwidth of 8 MHz. However, the frequency spectrum is too crowded to permit each station to occupy the full 8-MHz band for its picture signal transmission, hence the signal bandwidth is narrowed. In fact the FCC allows a total channel width of only 6 MHz for both picture and sound signals. This is accomplished by the use of vestigial sideband transmission.

Vestigial Sideband Transmission

The inclusion of both the picture and the sound carriers with their modulation components within a total channel bandwidth of 6 MHz is made possible by using *vestigial sideband transmission* techniques. This can be defined as transmission in which one sideband is partially suppressed while the other is transmitted unaltered.

Authorized Bandwidth of the
Transmitted Picture Signal

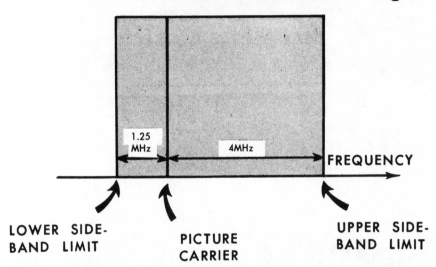

The completed upper sideband of 4 MHz is transmitted, but the lower sideband is transmitted at maximum amplitude over a relatively narrow frequency band of .75 MHz. From there, the progressively higher frequency components are gradually attenuated, reaching complete attenuation or zero signal level at 1.25 MHz. Thus all frequency components up to .75 MHz in the composite video signal are transmitted as double sideband; all frequency components between .75 MHz and 1.25 MHz become more and more upper sideband, and above 1.25 MHz are only upper sideband. This correction is the function of the vestigial sideband filter (which we show in the complete transmitter block diagram on the next page), or, in some instances, is accomplished in the power amplifiers by operating them off-tune. Thus the required picture-modulation components are confined to a total band-width of 5.25 MHz. This permits the picture and sound signal radiations from a transmitter to be within an overall signal bandwidth of 6 MHz.

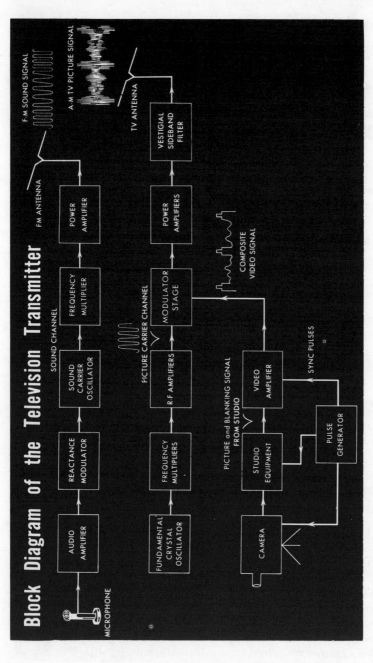

Block Diagram of the Television Transmitter

The Complete Television Transmitter

This description of a black and white television transmitter is a generalization, rather than the block diagram of any one particular transmitter. Numerous elements relating to control have been omitted. Only those details integral to the formation of the sound and video signals in the transmitter are illustrated.

Transmission Lines and Filters

The transmission line is the final link between the audio-video transmitters and the television antenna. Basically the transmission line is nothing more than a pair of wires or a cable that delivers to the antenna the power generated by the transmitters. The most common type of transmission line is the coaxial cable. It may consist of concentric conductors—a center wire passing through an outer cylindrically shaped wire braid, and a low-loss insulator keeping the wires correctly spaced. A more efficient line used in many high-power installations is a coaxial cable consisting of two coaxial cylindrical conductors insulated from each other by low-loss spacers. In this case the dielectric between the conductors can, for all practical purposes, be considered as air. In some instances, the space between the conductors is filled with an inert gas such as xenon, under pressure, to keep moisture out of the line and prevent oxidation.

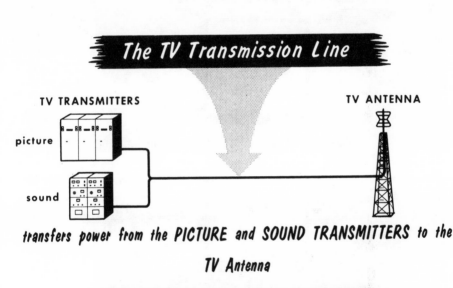

The TV Transmission Line

TV TRANSMITTERS   TV ANTENNA

picture

sound

transfers power from the PICTURE and SOUND TRANSMITTERS to the TV Antenna

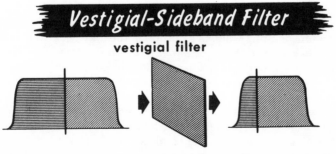

Vestigial-Sideband Filter

vestigial filter

removes portion of LOWER SIDEBAND of TV SIGNAL

Transmission Lines and Filters (cont'd)

Since it is common practice to transmit both sound and picture carriers from the same antenna, a special circuit called a *diplexer* is used. Into this unit is fed the RF signals from the sound and the picture transmitters. They are mixed and then fed to a common antenna from where they are radiated.

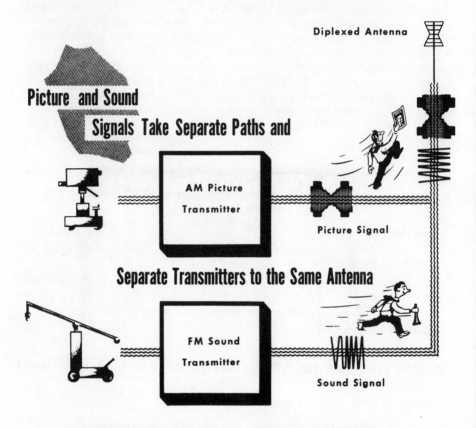

As will be discussed later in this course, a vestigial type of transmission is used in television broadcasting wherein a portion of the lower sideband or the transmitted signal is removed. Since essentially the same information is contained in both the upper and lower sidebands, considerable broadcasting spectrum space can be saved by using only one of the sidebands.

The circuit used to attenuate one of the sidebands is called a vestigial-sideband filter. It consists basically of a highly selective circuit tuned in the region of the unwanted sidebands. This filter can be placed in the final stage of the picture transmitter, or in the transmission line between the transmitter and the broadcast antenna.

Television Transmitting Antennas

The final step in the transmission of television signals is the radiation of the picture and sound carriers from the broadcast antenna. To send these signals the greatest possible distance, the transmitting antenna is usually located at the highest possible point in an area. Electromagnetic energy radiated from this antenna is directed toward the greatest concentration of television receivers. In some locations, the antenna is made omnidirectional (designed to radiate equally well in all directions).

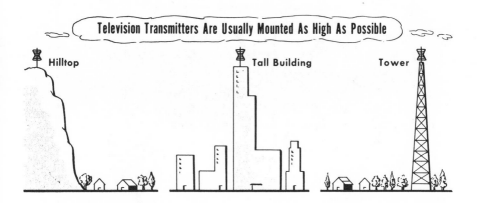

Television Transmitters Are Usually Mounted As High As Possible

Hilltop   Tall Building   Tower

To obtain this uniform signal distribution a *turnstile* antenna is used. To further concentrate the radiated picture and sound carriers in a horizontal plane about the broadcast antenna, the *superturnstile* antenna is used.

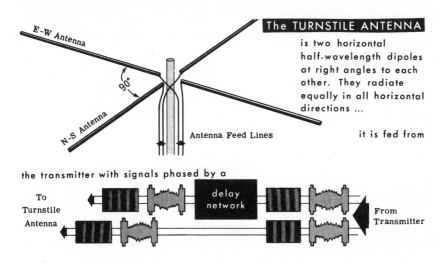

E-W Antenna

90°

N-S Antenna

Antenna Feed Lines

The TURNSTILE ANTENNA

is two horizontal half-wavelength dipoles at right angles to each other. They radiate equally in all horizontal directions ...

it is fed from

the transmitter with signals phased by a

To Turnstile Antenna

delay network

From Transmitter

Television Transmitting Antennas (cont'd)

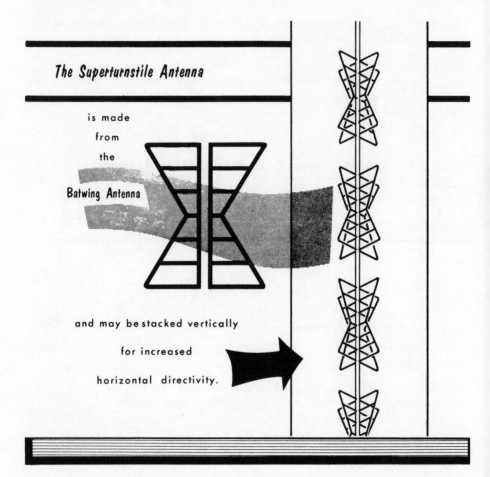

The Superturnstile Antenna

is made

from

the

Batwing Antenna

and may be stacked vertically

for increased

horizontal directivity.

This antenna is produced by the vertical stacking of *batwing* turnstiles. Each batwing unit consists of a series of horizontal elements or rods. The center rod is the shortest, giving the antenna the appearance of a batwing. This device may be considered as an extremely broad dipole antenna, useful for extending the frequency range of the broadcast antenna.

The peak output power of a television broadcast antenna may be as high as 50 kW in a typical high-power installation. However, depending upon the efficiency of the antenna, the effective radiated power (erp) may be many times greater.

Other antennas such as the *supergain* and the *helical* are also used in television transmission, but have not achieved the popularity of the superturnstile.

Television Standards in Other Parts of the World

The American television standards result in what has been generally agreed upon to be an acceptable picture. Notwithstanding, engineers in other countries have set their own standards substantially different from ours. Two respects in which they are the same are the interlace of the two-field transmissions, and the aspect ratio of 4:3. It might be mentioned that a number of years ago the British standard was a 5:4 aspect ratio. Because of the differences in the number of lines per field and per frame used in other systems, the horizontal and vertical sync as well as blanking intervals differ somewhat, although not necessarily too greatly, from our American standard.

It is possible to separate all of the other standards into three groups; European, British and French. All of the European countries do not necessarily use the European system; some use the British and some use the French. The European system involves the transmission of 625 lines per frame and 50 fields or 25 frames per second. The horizontal scanning frequency is 15,625 Hz. Whereas the American system allows from 483 to 499 active or unblanked horizontal scanning lines, the European system allows from 563 minimum to 589 maximum active or unblanked lines.

The British system is based on 405 lines per frame and 50 fields, or 25 frames per second. The horizontal scanning frequency is 10,125 Hz and calls for 377 active or unblanked lines per frame.

The French system involves 819 lines per frame and also 50 fields or 25 frames per second. Because of the greater number of lines per frame the horizontal frequency is naturally higher being 20,475 lines per second. The number of unblanked lines is 737.

While the sound portion of the television signal is frequency modulated in all of North and South America, this by no means holds true for the rest of the world. Many countries in Europe and Africa use amplitude modulation. Some areas, e.g., Vatican City, use amplitude modulation for one system and frequency modulation for another. In the American system of frequency modulation the maximum frequency deviation of the carrier is plus or minus 25 kHz, the European system makes use of a 50-kHz deviation corresponding to 100% modulation.

Another item of considerable variation is the television channel width. Here again all television broadcasting in North and South America uses a channel width of 6 MHz. The French system using 819 lines per frame requires a channel width of 14 MHz, while the other systems make use of channel widths of 6, 7, 8, and 9 MHz.

Television Standards In Other Parts of The the World

With the exception of Argentina and Venezuela, all nations in this hemisphere which enjoy television broadcasts use the same standards of operation as does the United States.

France and her colonies use two systems; 441 and 819 lines per frame, 25 and 50 frames per second. AM for audio transmission. 14 MHz bandwidth for each television channel

Great Britain uses 405 lines per frame, 25 frames per second. AM for transmission. 5 MHz bandwidth per channel.

The European system used in many countries is 625 lines per frame, 25 frames per second. AM and FM for audio transmission. According to standards in the individual countries, 5, 7, 8 and 14 MHz channel bandwidth.

Japan and the Philippines use American standards ; other countries use either the European or the French Systems.

Television Transmission

In some instances additional work is required in the distribution of the television signal. Some sparsely populated areas cannot provide financial support for a local television station, therefore a TV signal must be brought to them from some distant point. In other areas there may be only one station, and the large networks not represented in these areas must "bring in" their signals.

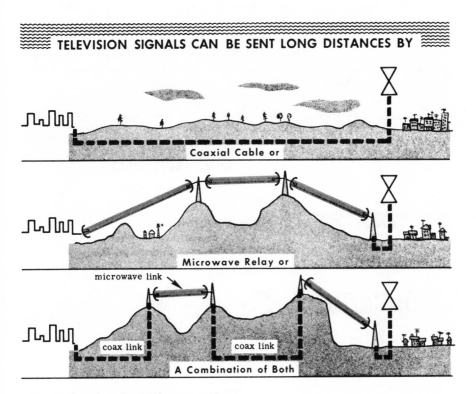

TELEVISION SIGNALS CAN BE SENT LONG DISTANCES BY

Coaxial Cable or

Microwave Relay or

microwave link

coax link   coax link

A Combination of Both

Two methods are used for *intercity transmission* of television signals. One is the use of long *coaxial cable transmission* lines that stretch for hundreds of miles. These lines, which are buried in the ground or strung along poles, contain amplifiers that periodically boost the signal along the line. The other popular method is *microwave transmission,* in which a series of microwave repeaters or relays are set 40 to 50 miles apart. Each relay receives the television signal and retransmits it. The radiation is in the form of a very narrow beam, and is *focused* by parabolic reflectors. When the signal reaches its final destination, its carrier-wave frequency is reduced to the required channel frequency and broadcast to the local community.

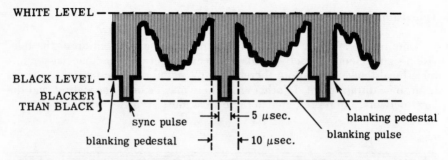

WHITE LEVEL

BLACK LEVEL

BLACKER } THAN BLACK }

sync pulse

blanking pedestal

5 μsec.

10 μsec.

blanking pedestal

blanking pulse

1. *Horizontal Sync Pulses* ride on horizontal blanking pulses, but they are easily "clipped" from the blanking pulses because they are shorter in interval. Sync pulses appear at the end of each horizontal scan to synchronize the information on the receiver screen with that on the camera tube.

2. *The Vertical Sync Pulses* allow horizontal syncing to be continuous, and also regulate interlacing.

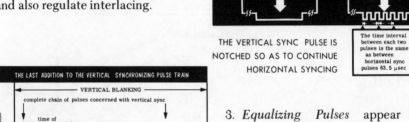

Popular or simplified version of the vertical sync pulse

Technical or actual version of the vertical sync pulse

THE VERTICAL SYNC PULSE IS NOTCHED SO AS TO CONTINUE HORIZONTAL SYNCING

The time interval between each two pulses is the same as between horizontal sync pulses 63.5 μsec

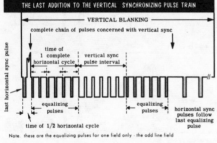

THE LAST ADDITION TO THE VERTICAL SYNCHRONIZING PULSE TRAIN

VERTICAL BLANKING

complete chain of pulses concerned with vertical sync

time of 1 complete horizontal cycle

vertical sync pulse interval

equalizing pulses

equalizing pulses

horizontal sync pulses follow last equalizing pulse

last horizontal sync pulse

time of 1/2 horizontal cycle

Note: these are the equalizing pulses for one field only - the odd line field

3. *Equalizing Pulses* appear in groups of six before and after each vertical sync pulse. They equalize the time difference created by the halflines of each field, and trigger the vertical retrace so that correct interlacing takes place.

4. *Picture-Carrier Generation.* The picture-carrier frequency is raised by a series of frequency multipliers; it is then passed through a number of amplifier stages; then, in the modulator stage, it is combined with the composite video signal.

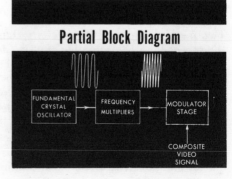

Partial Block Diagram

FUNDAMENTAL CRYSTAL OSCILLATOR

FREQUENCY MULTIPLIERS

MODULATOR STAGE

COMPOSITE VIDEO SIGNAL

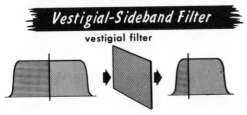

removes portion of LOWER SIDEBAND of TV SIGNAL

5. *Power Amplifiers* raise the modulated carrier to the level required for transfer to the antenna with the desired radiated power; they also limit the transmission of the modulated picture-carrier lower sidebands.

6. *Picture Signal Sidebands* transmit TV intelligence. The American system develops a total bandwidth of 8 MHz, but the FCC allows a total channel width of only 6 MHz. Therefore one sideband is partially suppressed by a vestigial sideband filter.

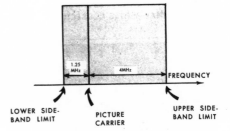

7. *Television Transmitting Antennas* which radiate the modulated picture and sound signals are of several types; one is the turnstile antenna which provides uniform signal distribution, another is the superturnstile which concentrates the radiated signals in a horizontal plane about the broadcast antenna.

8. *Television Transmission.* TV signals can be received from distant points by one of two methods: the use of hundreds-of-miles-long coaxial cable transmission lines, or a series of microwave repeaters or relays set 40 to 50 miles apart.

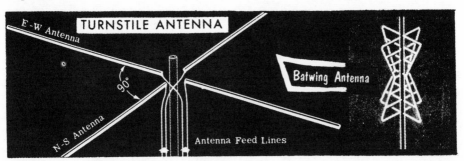

INDEX